Cities and Nature

For further volumes:
http://www.springer.com/series/10068

Dan Malkinson • Danny Czamanski •
Itzhak Benenson
Editors

Modeling of Land-Use and Ecological Dynamics

Springer

Editors
Dan Malkinson
Geography and Environmental Studies
and The Golan Research Institute
University of Haifa
Haifa, Israel

Danny Czamanski
Faculty of Architecture
and Town Planning
Technion-Israel Institute of Technology
Haifa, Israel

Itzhak Benenson
Department of Geography and
Human Environment
Tel-Aviv University
Tel Aviv, Israel

ISBN 978-3-662-51410-8 ISBN 978-3-642-40199-2 (eBook)
DOI 10.1007/978-3-642-40199-2
Springer Heidelberg New York Dordrecht London

Printed on acid-free paper

Springer is part of Springer Science+Business Media (www.springer.com)

Preface

Following the publication of our review paper concerned with the interaction of urban spatial evolution and ecosystems in 2008 (Czamanski et al. 2008) we received considerable reactions that included some criticism, doubts and much encouragement. We decided to start a major research project to place the various claims on a sound scientific basis. We were fortunate to receive funding from the Israel Science Foundation and from the government of Lower Saxony. Soon after, we signed contracts to publish two books in a new series to be published by Springer and entitled *Cities and Nature*. This edited volume is the first of the two books.

Our aim was to study the phenomena highlighted in our review paper and to present our results in one book. In particular, we set out to explore the interactions among the evolution of urban built areas, urban and peri-urban agriculture, open space networks and biodiversity. Among other activities, we decided to invite a group of scholars to address the same issues, to explore their findings and to publish them in an edited volume. In 2010 we were successful to obtain funding for an extended workshop to be held at the Technion – Israel Institute of Technology in early October 2011. For this we are grateful to the Azrieli Fund. The contributors to the edited volume were asked to present at the workshop their initial chapters. After the workshop the chapters underwent major revisions and review process. The chapters in this volume represent the fruits of this effort.

The participants in the workshops included 11 scholars from Europe and North America. Each participant presented their work and received extensive comments from other participants and the editors. The chapters were submitted and underwent another round of review, comments and corrections. We are grateful to the authors for their willingness to participate in the prolonged review and corrections process. We are certain that the results provide an interesting platform for continued study of the interactions of cities and nature and a worthy foil to our upcoming volume that explores the same issues as a result of our major research effort.

Reference

Czamanski, D., Benenson, I., Malkinson, D., Marinov, M., Roth, R., & Wittenberg, L. (2008). Urban dynamics and ecosystems. *International Review of Environmental and Resource Economics, 2*, 1–45.

Contents

Introduction

Danny Czamanski, Itzhak Benenson, and Dan Malkinson

The driving ideas at the backdrop of the current discussion of urbanization, sprawl and sustainability is the notion that urbanization is associated with low-density sprawl (Duany and Talen 2002; Sushinsky et al. 2013) and that sprawl reduces the amount of open spaces, fragments open spaces (Forman 1995, p. 418) and as a result adversely affects biodiversity (Fahrig 2001; Fahrig 2003; Alberti 2005; Donnelly and Marzluff 2006; Groom et al. 2006; Theobald et al. 2012). It is far from certain that these notions describe precisely the extant reality. While sprawl does reduce the amount of open space within boundaries of cities and does cause fragmentation, it does not necessarily reduce biodiversity. In some cities, the fragmented patches of open spaces remain interconnected allowing living spaces for plants and animals. Indeed, some view polycentric urban expansion as an opportunity to possible amelioration of declining biodiversity (Czamanski et al. 2008).

There is a growing public interest in the impact of urbanization on sustainability and in government actions to mitigate associated adverse repercussions on biodiversity. While much effort is being devoted to fashion policies, the relevant discourse and consequent policy prescriptions are broad, often vague and obscured by stylized facts and excessively broad conceptions of the relevant phenomena.

D. Czamanski (✉)
Faculty of Architecture and Town Planning, Technion-Israel Institute of Technology, Haifa, Israel
e-mail: danny@czamanski.com

I. Benenson
Department of Geography Human Environment, Tel-Aviv University, Tel Aviv, Israel
e-mail: bennya@post.tau.ac.il

D. Malkinson
Geography and Environmental Studies and The Golan Research Institute, University of Haifa, Haifa, Israel
e-mail: dmalk@geo.haifa.ac.il

D. Malkinson et al. (eds.), *Modeling of Land-Use and Ecological Dynamics*, Cities and Nature, DOI 10.1007/978-3-642-40199-2_1, © Springer-Verlag Berlin Heidelberg 2013

It is the purpose of the proposed book to shed light on the obscure and to place the discussion on a solid scientific basis. The contributors to this volume are an active group of scholars that span various disciplines, utilize advanced methods and make use of the vast, spatially detailed data that has become available in recent years. Their work models the relevant phenomena precisely and provides a basis for carefully fashioned alternate policies.

The major objectives of the proposed book are to:

- Describe conceptual models of the interactions among the three main types of land-uses: urban, agricultural and natural.
- Characterize the dynamics of city-agriculture-nature interfaces.
- Illustrate the conceptual models by means of case studies so as to reveal the particular forces and interactions that govern the respective interface dynamics.
- Develop and introduce land-use policies, planning measures and land-use planning tools to promote the sustainability of boundary areas.
- Assess the relationship between different taxa of species and the structure of the urban landscape.

The significance of these studies should be understood in the context of the growing urbanization of the world. In the middle of the second decade of the twenty-first century, over 50 % of the world's population is living in cities. This is remarkable since the process of urbanization that brought this about started in earnest less than 200 years ago. At that time, the number of people living in cities was only about 2 %. By 1900, the number of urbanites grew to 12 % (United Nations Population Division 2002). In other words, the last 100 years witnessed a huge human migration into cities. The slow clustering of people into small settlements during early biblical times, into villages and towns some 3,000 years ago and eventually into cities some 200 years ago has given rise now to a veritable flood. While forecasts of the world's future urban population are marred by many difficulties, including differences among countries in the definition of cities and urban areas as well as in forecasting methods, the United Nations has prepared such forecasts. Of the more than 2.2 billion in the world's population that will be added until the year 2030, some 95 % will live in cities (Cohen 2003). The world's population and economic activities are increasingly concentrated in space.

Scholarly interest in the uneven geographic evolution of the world's population and of economic activities started with von Thunen, also almost two centuries ago (Thünen 1826). It has been growing slowly, but steadily, ever since. A major impetus for this was the early work of the late Walter Isard (1956). But the scientific interest in urban phenomena peaked during the last decade of the twentieth century and the first years of the twenty-first century. An outburst of new models aiming to explain the births of cities and the dynamic processes governing their evolution, and the uneven geographic distribution of economic activities in general, started with a 1991 paper and culminated in the award of the Nobel Prize in economics to Paul Krugman in 2008 (Krugman 1991). Until Krugman the neoclassical explanations for the location of cities and increasing spatial concentration of activities and people was anchored in local resource endowments, the so-called first-nature. Cities formed where there were resources creating a competitive edge. Since Krugman, the emphasis shifted to explanations anchored in the relative locations of economic agents, the so-called second-nature.

While we are quite sophisticated in our modeling, the state of our understanding of the relevant phenomena remains far from satisfactory. Much of the existing theoretical infrastructure yields only partial insights concerning obvious and well-documented urban phenomena. Perhaps this is not surprising. The history of science is filled with examples of theories that despite insurmountable evidence to the contrary were and are believed to be true. Sometimes the untrue, maintained, theories do not have a worthy replacement. Facts that do not accord well with the accepted theory remain perplexing and are a source of embarrassment to some scientists. It is not understood fully why quite often such facts are ignored.[1] At times, they lead to periods of frantic search for new theories and a new, albeit a partial, explanation is forthcoming. Yet, often the sheer conservatism of the scientific community keeps the accepted and insufficient theory alive and prevents the nascent newcomer theory from taking its legitimate place.

Just as in the case of the spatial concentration of populations, the public discourse concerning nature and the welfare of ecological systems in the midst and at the boundaries of the built areas of cities is colored by unsubstantiated facts, popularly believed to be true, and theories with little empirical evidence to support them. This is particularly true in discussions concerning the resilience of networks of urban open spaces, their dynamics, the character of the ecological systems that populate these areas and forecasts concerning their future.

Scientists, particularly European ones, have begun exploring such issues over 100 years ago. Warren (1871) published a report on the floral composition in the urban parks of London and Kreuzpointner (1876) studied the flora of Munich. While the issues relating to ecological diversity in urban areas have interested researchers for many years, the crux of the questions addressed has changed. With the increasing pressure imposed by human populations on ecosystems in general, and specifically on urban ones, research has expanded from merely descriptive studies, to studies addressing theoretical questions pertaining to eco-urban systems, and progressively towards studies resulting in management and planning recommendations.

While earlier studies were descriptive by nature, towards the end of the twentieth century more attention has been dedicated to the relationship between urban patterns and species diversity. In contrast to expectations and beliefs many studies demonstrate that diversity of certain taxa can be as high, or even higher, within the city boundaries compared to surrounding natural areas (McKinney 2008). While from an accounting point of view this might be good news, this may not necessarily be the case from an ecological perspectives, as some of these species were artificially introduced by people. This is particularly true for flora species, as home gardens and city parks contribute many species to the urban species pool.

[1] Richard Thaler (2010), the father of behavioral economics is conducting a research project to illustrate and explain this phenomenon. See http://www.edge.org/3rd_culture/thaler10/thaler10_index.html.

This effect, however, cascades up the trophic levels as insects and vertebrates are attracted the heterogeneity and diversity of habitats generated within the urban landscape. Even predator abundance seems to be higher within the city, but the species which seem to be the mostly highly affected by urban development are the large bodied carnivores and birds of prey (Fischer et al. 2012).

Acknowledging the biological diversity in cities, and increasing awareness of them being in peril, has led to many initiatives that further map diversity patterns, protect them and valuate the importance of functioning ecosystems within the urban landscape. Individual cities, including New York City, Rio de Janeiro, Cape Town, Chicago, Melbourne, Jerusalem, countries, including Singapore, Portugal, the United Kingdom, Israel and others, and the United Nations have all set out to map the diversity and evaluate the ecological and economic importance of these systems. Thus, on the one hand, cities are growing and urban areas continue to sprawl, resulting in conversion of open spaces and agricultural lands into built environments, increasing pollution loads and potentially threatening the existence of species. On the other hand, awareness of importance of biodiversity and ecosystem function within the urban landscape have led to a better understanding of the services they provide and to different to mindful planning and growth of urban areas. For example, it has been demonstrated that urban open areas serve to filter air pollution, buffer and drain storm water, reduce noise, regulate micro-climate, provide recreational areas which benefit human welfare, and can serve as habitats for endangered species, as in the case of New York City and the Peregrine Falcon (Bolund and Hunhammar 1999; Kiviat and Johnson 2013).

Despite the growing availability of very detailed data and the shrinking cost of obtaining and managing them, there has been relatively little effort to date to expose and to map out a detailed picture of the relevant reality. The common theories are not confronted rigorously with empirical evidence. Yet, just like in the case of urban spatial dynamics, here as well there is already massive factual evidence that does not accord well with some of the accepted theories. In this volume, we seek to present evidence that cannot be ignored and point out theoretical constructs that should be abandoned. We propose and analyze dynamic processes that conform well to the evidence and can serve as a basis for productive management of the built and the natural environments.

A very good example of the incongruence of theory and facts is the most popular and most quoted model of urban spatial structure. It views cities as monocentric cones (Alonso 1964). Although the model is static, it implies that over time cities grow from the center outwards. As a result of natural increase of population and of immigration, the competition for accessibility leads to an outward expansion of the built areas. The implied wave of expansion is presumed not to leave any open spaces. Ecological systems and biodiversity are presumed to be jeopardized and that many species will not be able to survive. This is contrary to empirical evidence. It is enough to examine the available detailed footprints of built areas in a variety of cities to realize that open spaces are abundant and that even in compact cities there are open spaces that support the continued existence of species and communities (Sandstrom et al. 2006; Young et al. 2009).

At the broadest level, the lack of agreement between theories and evidence is due to the inconsistency of the spatial and temporal resolution at which the theoretical and empirical analyses are carried out. Theories generally refer to reality in broad-brush fashion and refer to crude, stylized facts. Almost exclusively the scholarly literature concerning cities and their environments are based on the behavior of typical agents and per capita, or per unit, indicators that focus on that which is common to various places and masks that which is particular to them. The use of such metrics misleads us to think that various phenomena display linear relationships to size and that there exist average dynamics common to all places. In reality, these robust macro relationships are the result of non-linear interactions in the underlying micro dynamic processes and local emergent processes. They display sub-linear and super-linear relationships (Bettencourt et. al. 2010). The focus on macro relationships at a crude spatial and temporal resolution does not allow us to identify the particularities of processes in the individual city and ecological system. There is a growing body of empirical evidence that suggests that spatial dynamics reflect scaling laws and are place-specific (Benguigui et al. 2000; Benguigui and Czamanski 2004). Empirical evidence is increasingly very detailed with great geographical and temporal coverage.

At the most common resolution, theories are not only too general to accommodate facts, most of the theoretical infrastructure is static. Reality is viewed by means of a single, or several, still pictures and accompanying story that attempts to bridge the gap. Rarely are we presented with a fully developed story that accounts for a video-like stream of pictures of social and natural processes. Such dynamic pictures reveal nonlinear evolutions of the built and natural systems and surprising twists and turns in their interactions. Indeed, these are complex systems of many intertwined organizational levels starting from microstructures and ending with macrostructures. They are historically dependent and at times they display emergent properties and self-organization.[2]

There are myriad of obstacles on the way to satisfying theories of the dynamic interactions of cities and nature. Perhaps the most formidable of these stems from the complex nature of the dynamic processes that needs to be considered. There is a huge variance in the characteristic time of the various elements that comprise the urban and natural systems. The dynamic processes take place in different time frames. During certain periods, the processes are fast and during other periods, they are slow. For example, were we to depict the life cycle of the processes of several such systems by means of logistic functions, along the time axis, the incidence of accelerating part of the curves and the saturations will not coincide. While one of the processes is accelerating, another is slowing down.

The chapters in this book touch on various aspects of the above issues. They combine empirical field studies, including field observations, analysis of remote sensing and GIS data, and high-resolution urban and regional models for getting a

[2] Noted early exception is the Cadillac model of Krugman in Fujita, Krugman and Venables (1999).

deep insight into the land pattern on the city-agriculture-nature interfaces and on the human abilities to plan and manage these evolving socio-ecological systems.

Koomen and Dekkers explore the potential of geospatial analysis to characterise land-use dynamics in the urban fringe and in particular focus on the impact of land-use policies in steering these developments. They formalize the open space preservation policies in the Netherlands, simulate the potential implications of proposed policy changes and investigate at what extent the policies are effective in limiting urbanization of the areas restricted for development.

Hatna and Bakker employ a set of digitized historical land-cover maps in order to compare the spatial distribution of cropland, pasture, and nature surrounding cities in the Netherlands over a long time period – 1900, 1960 and 1990. While they find general stability in the land cover around cities, the growth in the amount of croplands near the perimeter of cities in 1900, was weakened by the middle of the century and almost completely ceased by 1990.

The chpater by Marceau, Wang, and Wijesekara describes the application of two cellular automata (CA) designed at two spatial scales to investigate the land-use dynamics occurring respectively in the whole watershed and in the eastern portion of the watershed, immediately adjacent to the City of Calgary. The first model is at a spatial resolution of 60 m. It provides information about alternative spatial distributions of urban areas that can occur according to spatial constraints imposed on land development. The second model is at a resolution of 5 m. It is a patch-based model. The models were used to simulate land development scenarios over a period of 20–30 years. The analyses reveal how land consumption can be considerably diminished by encouraging the protection of sensitive areas and increasing the density of existing and new urban residential areas.

Miguel Serra and Paulo Pinho investigate formation of suburban street networks at the urban fringe by studying Oporto dynamics over a period of 55 years. Their scrupulous high-resolution study employs space syntax approach identifies individual development operations and structural evolution of the entire street networks through simple topological parameters. The chapter demonstrates essential influence of the local developments on the developing network and offers a view of the bottom-up regulation of the street network development that can be translated in planning procedures.

The chapter by Matthies et al. compares between two urban landscapes, Hannover and Haifa, in an attempt to identify the factors driving vegetation diversity patterns within the open space patches s of the cities. Patch sizes and distances from the urban border were used as explanatory variables to predict species richness, native species richness and the proportion of native species within a patch. In spite of the fact that cities are located in different geographical and climatic areas, in both urban landscapes only patch size was found to be a significant factor dictating total species richness and native species richness.

The chapter by Runfola, Polsky, Giner, Pontius Jr., and Nicolson provides a study of the growth of lawns in the United States. Because lawns are maintained through fertilization and watering they present risks for water use and quality, nutrient cycling, urban climate regimes, and even human health. The associated ecological ramifications, such as habitat fragmentation, water quality and availability may be far-reaching. The authors produce a high resolution (0.5 m) land-cover classification to quantify existing lawn extent for the year 2005 in the Plum Island Ecosystem (PIE), a collection of 26 suburban towns northeast of Boston, MA, USA. They then use this dataset in conjunction with the GEOMOD land-change model to project lawn extent for the year 2030.

The chapter by Felsenstein, Lichter, Ashbel and Grinberger present a high-resolution model of the land use-land cover dynamics at the urban fringe that focuses on the reciprocal dependencies between the land cover and land use. They employ historical data on the land use and land cover in Tel Aviv metropolitan during last two decades, and explicitly simulate dynamics on the metropolitan fringe as dependent on the rate of population growth and land demand to the year 2023. When coupled with an appropriate biodiversity model, the land use-land cover model could be extended to forecasting the environmental stress of metropolitan expansion.

Insarov's and Insarova's chapter considers the effect of urbanization on lichens and vegetation. It focuses on the main processes taking place in urban lichens and plants at the level of organism, community and ecosystem, and the ecological services they provide. Species richness and composition, community and ecological processes, growth, anatomical, morphological traits and physiological processes under urbanization stress are discussed. Various ecological services provided by urban vegetation, which include: air quality improvement and abatement of noise, among other, are described. A number of case studies from various cities around, which assessed changes of urban lichen communities, are reported. The importance of lichens as monitors of temporal and spatial trends in the state of urban biota under air pollution stress is detailed, and implications for city planning and management are provided.

References

Alberti, M. (2005). The effects of urban patterns on ecosystem function. *International Regional Science Review, 28*(2), 168–192.

Alonso, W. (1964). *Location and land use*. Cambridge: Harvard University Press.

Benguigui, L., & Czamanski, D. (2004). Simulation analysis of the fractality of cities. *Geographical Analysis, 36*(1), 69–84.

Benguigui, L., Czamanski, D., Marinov, M., & Portugali, Y. (2000). When and where is a city fractal? *Environment & Planning B, 27*, 507–519.

Bettencourt, L. M. A., Lobo, J., Strumsky, D., & West, G. B. (2010). Urban scaling and its deviations: Revealing the structure of wealth, innovation and crime across cities. *PLoS One, 5*, e13541.

Bolund, P., & Hunhammar, S. (1999). Ecosystem services in urban areas. *Ecological Economics, 29*, 293–301.

Cohen, J. E. (2003). Human population: The next half century. *Science, 302*(5648), 1172–1175.

Czamanski, D., Benenson, I., Malkinson, D., Marinov, M., Roth, R., & Wittenberg, L. (2008). Urban dynamics and ecosystems. *International Review of Environmental and Resource Economics, 2*, 1–45.

Donnelly, R., & Marzluff, J. M. (2006). Relative importance of habitat quantity, structure, and spatial pattern to birds in urbanizing environments. *Urban Ecosystems, 9*(2), 99–117.
Duany, A., & Talen, E. (2002). Transect planning. *Journal of the American Planning Association, 68*(3), 245–266.
Fahrig, L. (2001). How much habitat is enough? *Biological Conservation, 100*, 65–74
Fahrig, L. (2003). Effects of habitat fragmentation on biodiversity. *Annual Review of Ecology Evolution and Systematics, 34*, 487–515.
Fischer, J. D., Cleeton, S. H., Lyons, T. P., & Miller, J. R. (2012). Urbanization and the predation paradox: The role of trophic dynamics in structuring vertebrate communities. *BioScience, 62*, 809–818.
Forman, R. T. (1995). *Land mosaics: The ecology of landscapes and regions*. Cambridge: Cambridge University Press.
Fujita, M., Krugman, P., & Venables, A. J. (1999). *The spatial economy*. Cambridge: MIT Press.
Groom, M. J., Meffe, G. K., & Carroll, C. R. (2006). *Principles of conservation biology*. Sunderland: Sinauer Associates.
Isard, W. (1956). *Location and space-economy; a general theory relating to industrial location, market areas, land use, trade, and urban structure*. Cambridge: Wiley.
Kiviat, E., & Johnson, E. A. (2013). Biodiversity assessment handbook for New York City. American Museum of Natural History, Center for Biodiversity and Conservation, New York and Hudsonia Ltd., Annandale.
Kreuzpointner, J. B. (1876). Notizen zur Flora Munchens. *Flora, 59*, 77–80.
Krugman, P. (1991). Increasing returns and economic geography. *Journal of Political Economy, 99*(3), 483–499.
McKinney, M. L. (2008). Effects of urbanization on species richness: A review of plants and animals. *Urban Ecosystems, 11*, 161–176.
Sandström, U. G., Angelstam, P., & Mikusiński, G. (2006). Ecological diversity of birds in relation to the structure of urban green space. *Landscape and Urban Planning, 77*(1), 39–53.
Sushinsky, J. R., Rhodes, J. R., Possingham, H. P., Gill, T. K., & Fuller, F. R. (2013). How should we grow cities to minimize their biodiversity impacts? *Global Change Biology, 19*(2), 401–410.
Thaler, R. (2010). The father of behavioral economics is conducting a research project to illustrate and explain this phenomenon. http://www.edge.org/3rd_culture/thaler10/thaler10_index.html
Theobald, D. M., Reed, S. E., Fields, K., & Soule, M. (2012). Connecting natural landscapes using a landscape permeability model to prioritize conservation activities in the United States. *Conservation Letters 0*, 1–11.
Thünen, J. H. (1826). *Der Isolierte Staat in Beziehung Auf Landwirtschaft Und Nationalökonomie*. Hamburg: Perthes. German: Wartenberg, C.M. (1966) *Von Thünen's Isolated State* (trans: Wartenberg, C.M.). Oxford: Pergamon Press.
United Nations Population Division. (2002). *World urbanization prospects: The 2001 revision*. ESA/P/WP.173. New York: United Nations.
Warren, J. B. L. (1871). *The flora of Hyde Park and Kensington Gardens*. London: Taylor and Co.
Young, C., et al. (2009). Urban landscape ecology and its evaluation: A review. In A. Dupont & H. Jacobs (Eds.), *Landscape ecology research trends* (pp. 45–69). New York: Nova Science.

The Impact of Land-Use Policy on Urban Fringe Dynamics

Dutch Evidence and Prospects

Eric Koomen and Jasper Dekkers

Abstract Concern for the loss of open space around urban areas has given rise to various forms of land-use policy that aim to steer urban fringe dynamics. This chapter explores the potential of geospatial analysis to characterise land-use dynamics in the urban fringe and in particular focuses on the impact of land-use policies in steering these developments. The Netherlands is used as a case study because this country has a long-standing tradition of applying such polices and is generally considered to represent a successful example of restrictive spatial planning. Yet, these policies have received substantial criticism in the past decade and are currently being transformed by the National Government. Based on the observed degree of success of current open space preservation policies we make an attempt to simulate the potential implications of the proposed policy changes.

1 Introduction

1.1 Land-Use Policy and Urban Dynamics

Urban fringe dynamics put pressure on the open spaces that surround urban areas and thus limit the potential of these green areas for biodiversity, agricultural production and a wide range of other landscape services such as: water regulation and storage, air quality improvement and recreational opportunities. Concern for the deterioration of these services has given rise to various forms of land-use policy that aim to steer urban fringe dynamics. Open space preservation policies in western countries generally aim to manage urban growth through a range of concepts such as: zoning, urban growth boundaries, transfer of development rights

E. Koomen (✉) • J. Dekkers
Department of Spatial Economics/SPINlab, VU University Amsterdam, De Boelelaan 1105, 1081 HV Amsterdam, The Netherlands
e-mail: e.koomen@vu.nl; j.dekkers@vu.nl

D. Malkinson et al. (eds.), *Modeling of Land-Use and Ecological Dynamics*, Cities and Nature, DOI 10.1007/978-3-642-40199-2_2, © Springer-Verlag Berlin Heidelberg 2013

and related financial instruments (Frenkel 2004). These policies relate to the contested topic of urban sprawl and are a key planning theme in the United States and many other western countries (Bartlett et al. 2000; Romero 2003; Bae and Richardson 2004; Gailing 2005). A useful overview of common approaches and methods in open-space planning is provided by Maruani and Amit-Cohen (2007). Van der Valk and Van Dijk (2009) present an extensive discussion of planning options and their relation with actual open space dynamics and planning institutions.

This chapter explores the potential of geospatial analysis to characterise land-use dynamics in the urban fringe and in particular focuses on the impact of restrictive land-use zoning policies in steering these developments. As a case study we selected the Randstad region in the Netherlands. This is a typical urban fringe region consisting of a mix of big cities, towns and villages surrounded by a substantial amount of open space as will be described in the next section. From a policy perspective, the Netherlands offers an interesting case study environment because it is generally considered to have a successful institutional policy framework for the management of urban development and the preservation of open space (Alpkokin 2012; van der Valk 2002; Alterman 1997; Dieleman et al. 1999; Koomen et al. 2008b). Yet, several studies also mention negative impacts that can be associated with the implemented land-use policies in the Netherlands. These impacts include issues such as: the mismatch between location of homes and jobs that leads to an increase in car use (Schwanen et al. 2001); the sharp increase in land prices following from the combination of spatially explicit zoning regulations and the deregulation of housing development at the end of the last century (Dieleman et al. 1999); and the limited fulfilment of the demand for green types of suburban housing (Rietveld 2001). These issues gave rise to an ongoing debate about the future of spatial planning in the Netherlands (e.g. Korthals Altes 2009; Roodbol-Mekkes et al. 2012). Following this debate National Government has recently introduced new, less stringent types of restrictive policies and is currently considering abolishing the more restrictive policies that were typical for the Netherlands.

We aim to assess the relative effectiveness of the various restrictive land-use policies that were in place in the past 15 years and demonstrate how this knowledge can be used to assess the potential impact of proposed policy changes. The presented results are of particular interest to other countries that consider the introduction or revision of similar land-use policies.

1.2 The Dutch Randstad in Perspective

The Netherlands is often considered as one of the most densely populated countries in the world. While this popular image is close to the truth – the country's population density of 400 inhabitants per km^2 rank it in the top ten of independent countries[1] – it is important to consider that the major Dutch cities are relatively

[1] Source: United Nations (2011).

small. The two biggest cities of the country (Amsterdam and Rotterdam) each have a total population of less than one million inhabitants and do not even qualify for a position in the top 500 of largest cities in the world.[2]

In fact, the urban system in the most urbanised western part of the Netherlands has a peculiar layout: it consists of four larger cities and several smaller ones that are separated from each other by open spaces and that together form a horseshoe-shaped settlement pattern around a central open space that is mainly used for agriculture. This layout was first recognized in the 1920s (Faludi and Van der Valk 1994) and has since then become known as the combination of the Randstad (literally rim-city) and Green Heart. These elements have been central to Dutch spatial planning in the Netherlands in past 50 years and their origin, characteristics and evolution has been described in many publications (Burke 1966; Hall 1966; Ottens 1979; Faludi and Van der Valk 1994). To this date the elements of this specific layout are recognisable: the Randstad and Green Heart now contain about 6.5 million inhabitants, but the differences between the cities and surrounding open areas are substantial. The individual cities have densities of, for example, 6,500 (The Hague) or 4,500 (Amsterdam) inhabitants per km^2 and are among the most densely populated urban areas of Europe.[3] The density in the Green Heart on the other hand is about 10 times lower (595 inhabitants per km^2).[4]

An initial idea about the magnitude of land-use change in the country can be obtained from the aggregate national statistics provided by Statistics Netherlands. Figure 1 shows how the amounts of land devoted to the main types of use have changed in the past century. It clearly indicates that the Dutch landscape is still dominated by agriculture. In the first half of the twentieth century the amount of agriculture even increased as a result of the conversion of open nature areas (mainly heath land) and the reclamation of large inland water bodies (not included in the presented statistics). From about 1970 onwards the amount of agricultural is decreasing, mainly to accommodate the strong growth of urban areas and, to a lesser extent, the increase in forest following the national plans for the development of a National Ecological Network. An extensive discussion and international comparison of this planning concept is provided elsewhere (Jongman et al. 2004).

These aggregate statistics sketch the general, national trends in spatial developments but cannot be used to assess the impact of land-use policies. Such policies do not necessarily aim to change national developments, but rather steer local land-use changes. These changes and their relation with restrictive policies will be analysed with different methods and datasets as is discussed in the next section.

[2] Source: www.citymayors.com.

[3] According to the Urban Audit of Eurostat (http://epp.eurostat.ec.europa.eu).

[4] This density is based on data from Statistics Netherlands (2012) and calculated by dividing the total number of inhabitants of all municipalities that fall for more than 50 % within the Green Heart by their land area.

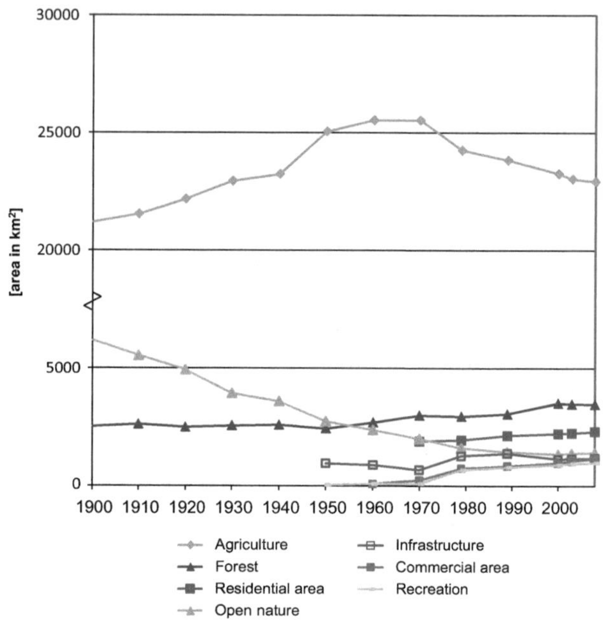

Fig. 1 Historic land-use changes in the Netherlands 1900–2008 (Source: Statistics Netherlands 2012). Note that some types of land use are only registered in the second part of the century

1.3 Longstanding Planning Tradition Under Revision

Although Dutch national spatial planning has its roots in pre-war concepts and ideas (Faludi and Van der Valk 1994), it only came to fruition in 1958 when the main principles for the development of the western part of the Netherlands were presented by the State Service for the National Plan (RNP 1958). Their report proposed to steer the expected strong urbanisation pressure away from the Green Heart and the existing cities in the Randstad to growth centres just outside the rim of cities (Fig. 2). This concept had several objectives: accommodating the expected urban growth in the direct vicinity of the Randstad, preserve the agricultural Green Heart, and prevent the existing cities from growing together and becoming metropolises with more than one million inhabitants (Faludi and Van der Valk 1994).

These main ideas were central to the five national planning reports that were drafted by the Ministry responsible for public housing and spatial planning in the following 50 years and that formed the basis for planning at lower tiers of Government. Key elements in these reports were: the preservation of the central open space and the designation of eight Buffer zones[5] between the major cities of the Randstad (V&B 1960) and bundled deconcentration (V&RO 1966). This latter policy initiative resulted in the development of New towns were suburbanisation could be

[5] These are strategically designated green corridors between large urban areas that aim to prevent them from growing together.

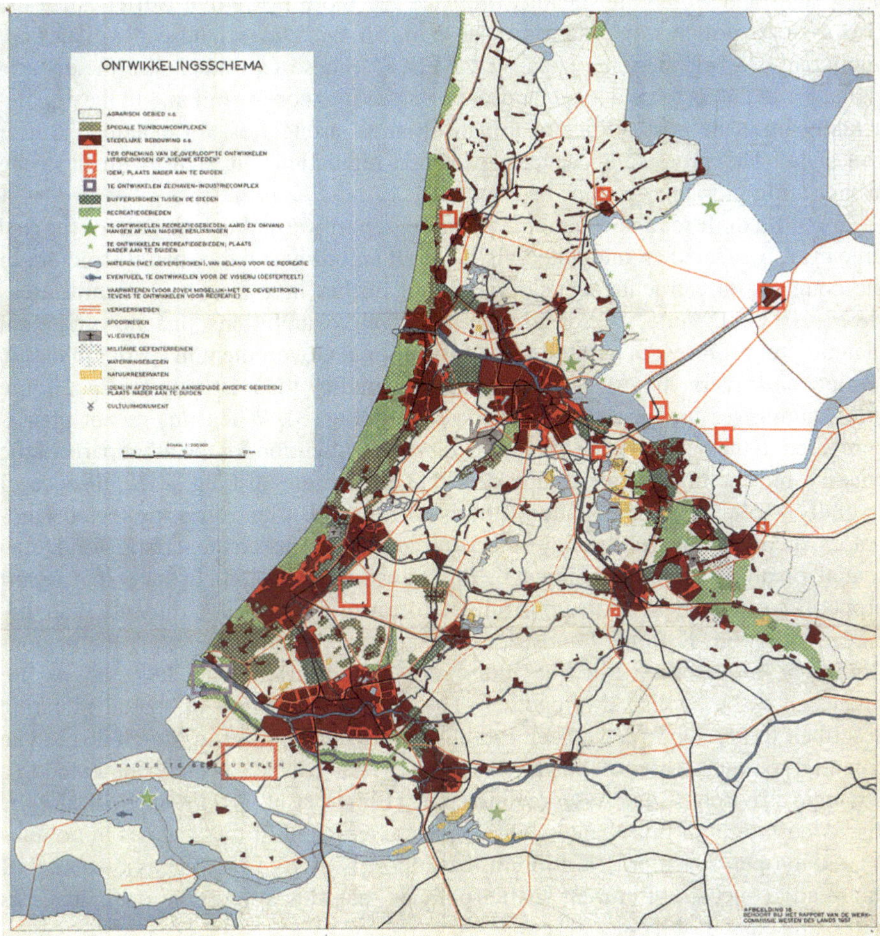

Fig. 2 The plan for the development of the western part of the Netherlands (Source: RNP 1958). *Squares* denote to be developed New towns, hatched areas are Buffer zones or recreation areas

concentrated. Following the observed population decline in the big cities, the strong increase in commuting and related car use and a growing awareness for landscape preservation, subsequent planning reports aimed to bring about more compact cities by concentrating urbanisation in the vicinity of existing cities in combination with restrictive policies on open areas (V&RO 1977; VROM 1989). The latest approved planning report (VROM et al. 2004) maintains these principles but offers a more liberal view on planning and shifts its attention from restriction of urban development in protected areas towards stimulation: a change from 'no, unless' to 'yes, if'. Regional and local governments, private organisations and enterprises are now provided with more freedom to meet their objectives. This report also paid explicit attention to the cultural historic values of the 20 National Landscapes that it designated and which are meant to replace the more restrictive Buffer zone and Green Heart policies, as well as those parts of the Dutch landscapes that were approved as world heritage sites by UNESCO. These two labels are meant as a signal for planners to take local landscape values into consideration, but they do not actually limit the possibilities of urbanisation as strongly as, for example, Buffer zones do. The protective capacity of these regulations in areas with a strong urbanisation pressure can thus be doubted.

In 2011 a draft for a new national planning report was published that, for the first time ever, presented an integrated vision of all ambitions on the spatial structure of the country, infrastructure development and further investments in natural areas (I&M 2011). Despite its considerable integrative ambitions this new national planning report foresees a limited role for National Government in spatial planning (Kuiper and Evers 2011b). This is in part reflected by the relatively abstract figure that summarises the spatial plan of the report (Fig. 3). When this recent plan is compared to the one of 1958 (Fig. 2) the increased attention for the other parts of the country and the hinterland are apparent, but even more striking is the absence of spatially explicit policies. None of the New towns, Buffer zones, protected landscapes or other concrete policy interventions that were characteristic for Dutch spatial planning are incorporated in this new spatial vision. In fact, this report proposes to transfer more responsibilities to lower tiers of Governments than any preceding report. National Landscapes are suggested to become solely the responsibility of provinces, while the more restrictive Buffer zone designations are suggested to be abolished altogether. The proposed changes in spatial planning described in this report have been established by the Minister in March 2012. The potential spatial impacts of the proposed policy changes have been documented in a Strategic Environmental Assessment report (Elings et al. 2011) and will also be discussed briefly in this chapter. The impact assessment of the changes in policies related to open space preservation in that report is partly based on assessments of the relative success of current spatial policies that aim to preserve open space as will be discussed in the next sections.

2 Methodology

This chapter describes the major land-use transitions in the past 15 years in the Randstad area based on a variety of highly detailed spatial data sources. Specific attention is paid to the impact of restrictive zoning policies in limiting urban development in specific areas. In this section we introduce the methods that we applied and spatial data sets, while the subsequent section will discuss the main results.

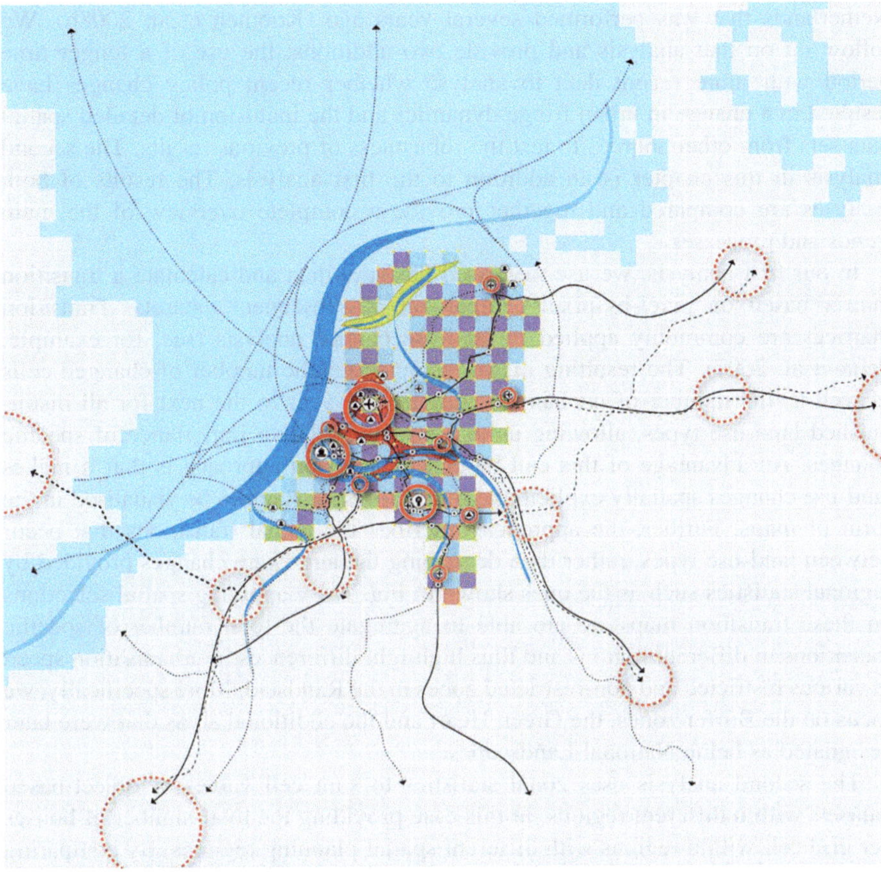

Fig. 3 The spatial vision depicted in the current established national spatial plan for the Netherlands (Source: I&M 2011). The *circles* denote core economic areas, the *coloured squares* symbolise existing mixes of land use, the *lines* depict important transport routes

2.1 Spatial Analysis Techniques

Spatial analysis techniques applied to detailed geographical data offer a useful tool to analyse the impact of spatial policies. Longley and others already used this combination in 1992 to assess the influence of the well-known Green Belt zoning regulations in Britain (Longley et al. 1992) by comparing the geometry of settlements which were subject to such policy with those which were not. Similar more recent examples of the spatial analysis of the impact of zoning regulations exist for, for example, Israel (Frenkel 2004), the Netherlands (Koomen et al. 2008b) and China (Zhao 2011).

In a first analysis, this chapter builds upon a quantitative GIS-based study into the relative success of the long-standing restrictive spatial policies in the

Netherlands that was performed several years ago (Koomen et al. 2008b). We follow up on that analysis and provide two additions: the use of a longer time period with more recent data to analyse whether recent policy changes have resulted in a change in urban fringe dynamics and the inclusion of detailed spatial data sets from other sources to test the robustness of previous results. The second analysis in this chapter is an addition to the first analysis. The results of both analyses are compared and together provide a complete overview of the main trends and processes.

In our first analysis, we use rasterised land-use data and calculate a transition matrix based on pixel-by-pixel comparison of subsequent datasets. Transition matrices are commonly applied in land-use change analysis (see, for example, Pena et al. 2007). The resulting matrix summarizes the number of changed cells as well as the number of unchanged cells from 1 year to the next for all distinguished land-use types, allowing us to assess the relative importance of specific changes. An advantage of this cell-by-cell comparison approach is that it makes land-use changes spatially explicit and thus allows for them to be visualised in the form of maps. Further, the approach describes the actual transitions that occur between land-use types, rather than describing the aggregate changes provided by regional statistics such as the ones shown in Fig. 1. By applying spatial selections on these transition maps we are able to aggregate the total number of specific transitions in different regions and thus highlight differences in urbanisation speed in various restricted and non-restricted zones in the Randstad. More specifically, we focus on the Buffer zones, the Green Heart and the additional areas that were later designated as being National Landscapes.

The second analysis uses zonal statistics to sum cell values of object-based datasets within different regions, in this case providing the total number of houses per grid cell within regions with different spatial planning regimes. By comparing the total number of houses on an aggregate level for different years, we can calculate growth in housing stock in absolute and relative terms. Zonal statistics can be created with standard GIS procedures that are available in most GIS-software packages.

2.2 *Land-Use Simulation*

Land-use change models are useful tools to support the analysis of the consequences of land-use change (Koomen et al. 2008a). They can, for example, help formulate adequate spatial policies by simulating potential autonomous spatial developments or, perhaps more importantly, by showing the possible consequences of different policy alternatives. Policy makers can thus be confronted with a context of future conditions and an indication of the impact the spatially relevant policies they propose.

The model we apply here – Land Use Scanner – is rooted in economic theory. It is an integrated land-use model that offers a view of all types of land use, dealing with urban, natural and agricultural functions. It has been developed in 1997 by a group of research institutes and has been applied in a large number of policy-related research projects in the Netherlands and abroad. The model's basics and recent calibration have been described extensively elsewhere (Koomen and Borsboom-van Beurden 2011; Loonen and Koomen 2009).

The model is often applied to perform what-if type of applications that visualise the spatial developments given development scenarios. In that respect it is comparable to well-known rule based simulation models such as the original California Urban Futures (CUF) model and the What If? system (Landis 1994; Klosterman 1999). In the context of strategic, scenario-based national planning, the model proved to be a especially valuable tool to inform policy makers about potential future developments (Schotten et al. 2001; Borsboom-van Beurden et al. 2007; Dekkers and Koomen 2007) or to provide ex-ante evaluations of policy alternatives in both national (Scholten et al. 1999; Van der Hoeven et al. 2009) and regional contexts (Koomen et al. 2011b; Jacobs et al. 2011).

The Land Use Scanner balances the demand and supply of land using three main model components:

1. External regional projections of land-use change, usually referred to as demand or claims, that are land-use type specific and can be derived from, for example, sector-specific models of specialised institutes;
2. A local (cell-based) definition of suitability that incorporates a large number of spatial datasets referring to current land use, physical properties, operative policies and market forces generally expressed in distance relations to nearby land-use functions;
3. An algorithm that allocates equal units of land (cells) to those land-use types that have the highest suitability, taking into account the regional land-use claim. This discrete allocation problem is solved through a form of linear programming (Koomen et al. 2011a).

2.3 Data Sets Used

2.3.1 Raster Data

We use four 25 m resolution versions of the spatial land-use databases from Statistics Netherlands referring to the years 1996, 2000, 2003 and 2008. These data sets allow the identification of main land-use change trends, despite some methodological issues: to a certain extent, classification is done by hand and there are some interpretation and definition differences from year to year (van Leeuwen 2004; CBS 2008). Part of the observed transitions will thus refer to differences in spatial delineation (i.e. interpretation and definition differences) from 1 year to the next.

The original datasets are aggregated into a limited number of major land-use classes that are necessary for analysing the urbanisation process. Seven main types of land use are distinguished based on the initial set of 38 land-use types: (1) built-up areas, (2) urban green (e.g. bare soil, parks and building lots with a functional relation to the neighbouring urban area), (3) greenhouses, (4) other agriculture, (5) infrastructure, (6) nature and (7) water. These aggregate classes allow for an effective analysis of the main land-use transitions in the studied period. The reason why we distinguish greenhouses separately from other agriculture is because of its distinct urban appearance and relevance to spatial planning. The latter is indicated by the fact that Dutch government has projects in place that aim to remove dispersed greenhouse locations from areas with a high landscape value such as the Buffer zones (see, for example, EL&I 2009). Apparently greenhouses are considered an unwanted form of land-use change.

The various transitions that occur between the main groups of land use can be captured in four land-use change processes: (1) actual urbanisation, (2) potential urbanisation, (3) nature development (i.e. construction of new nature, for instance as part of the National Ecological Network, NEN) and (4) other changes (e.g. infrequent transitions between agriculture and water, or between greenhouses and infrastructure). As our focus is on the main objective of land-use planning – limiting the conversion of open space into urban use- we disregard minor land-use modifications or changes in land-use intensity in our analysis.

2.3.2 Object Data

For the object-based analysis, we use a dataset containing the number of houses per 100 m grid cell for the years 2000, 2003, 2004 and 2008, provided by PBL Netherlands Environmental Assessment Agency. As Evers et al. (2005) explain, this dataset is based on a combination of data sets containing information on the location of houses and population. For our analysis, two underlying data sets are important: (1) ACN (*Adres Coördinatenbestand Nederland*) provided by the Dutch Land Registry Office (*Kadaster*) that contains the x,y-coordinates of all addresses in the Netherlands; and (2) 'Geo-Marktprofiel', a direct marketing data set from Bisnode containing a wide variety of data on characteristics of households and number of houses per six-digit zip code centroid in the Netherlands. Other datasets are used as well to determine the main function of objects, but these are less relevant so we do not discuss them here.

Noteworthy is the fact that the providers of these data sets have constructed them for their own specific purposes and not necessarily for the analysis of (spatial) changes in the number of houses and people. Next, the ACN-data set is known to be not entirely up-to-date in areas where a lot of address mutations take place, such as large-scale urban development locations and inner city reconstruction zones. Also, in rural areas the x,y-coordinate of an address may lie outside the objects (houses) it refers to but within the wider boundaries of the cadastral parcel. Potentially this means that the location of an object is misplaced by 50 m or more, possibly causing it to be added to an adjacent grid cell in the data set we use for our analysis. Further, the Geo-Marktprofiel data set is known to differ from the data of Statistics

Netherlands in terms of number of houses and households. This is caused by a different definition of the objects at hand. Statistics Netherlands, for example, does not consider house boats to be houses, where GeoMarktprofiel does. Validation analysis by other researchers who use the same data, however, indicates that these differences are very small and that the data corresponds to reality (see, for example, Tesser et al. 1995).

The most important data quality issue for our analysis, however, is the fact that before 2004, the ACN data set overestimated the number of addresses. This was due to an inconsistent registration of house number additions (e.g. 2 and II for second floor apartments) that in some cases made the same residential object show up twice in the data set (Evers et al. 2005). In 2004, a one-time correction has taken place, removing around 200,000 addresses from the dataset. Because of this fact, we are forced to split our analysis in two periods: 2000–2003 and 2004–2008.

3 Results

3.1 Raster-Based Analysis

This section describes the results of the raster-based analysis of land-use transitions between 1996 and 2008. In our analysis we focus on different restrictive regimes in the Randstad area: Buffer zones, the Green Heart and the newly formed, less restrictive National Landscapes. For the latter category we only look at those areas that were not previously protected under the more restrictive Buffer zones and/or Green Heart policies. Urban development in the restrictive policy zones is compared to developments in the non-restricted part of the Randstad.

Figure 4 provides an overview of the described zones and shows the urban development locations observed during 1996–2008. The figure indicates that most development takes place in relatively large concentrated urban extensions in the direct vicinity of the larger urban areas. These developments follow from the compact city philosophy that was especially prominent in the fourth national spatial planning report (VROM 1989). Most of these developments are located outside the restrictive policy zones. However, in some cases they lie within them, eventually leading to the adjustment of the zoning regulations. This is particularly prominent in the Green Heart zone that has been adjusted in 1993 and again in 2004 (see Pieterse et al. 2005; Koomen et al. 2008b). Such adjustments show the tension between policies that have a shared objective (open space preservation) but differ in their approach (promoting compact cities or regionally limiting urban development). The insets of Fig. 4 illustrate this process.

Tables 1 and 2 present the *net* land-use change for the most important transition processes in the three subsequent 3–5 year periods, aggregated for different spatial planning regions. They summarise all observed transitions between the seven aggregate land-use classes under four main processes of land-use change.

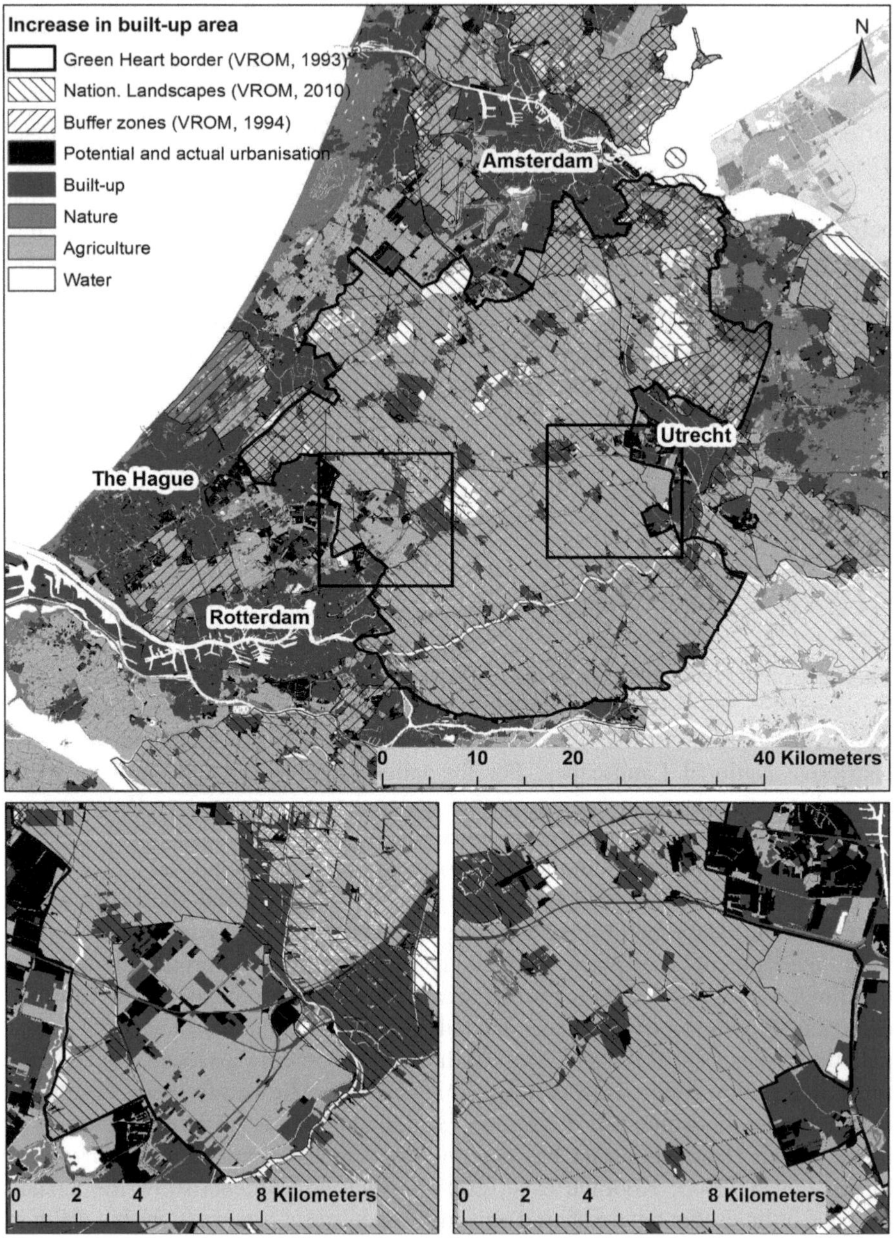

Fig. 4 Urban development in the Randstad (*highlighted area*) in the 1996–2008 period. The black locations depict all transitions into Urban area, Infrastructure, Urban green and Greenhouses and this includes both the potential and actual urbanisation processes described in the text. The insets show locations where the Green Heart contour has been adjusted and urbanisation has followed (*left*) or is likely to do so in the future (*right*)

Table 1 Most important net land-use transitions for the Buffer zones (within the Randstad only) and the Green Heart

Land use change (Net)		Buffer zones							Green Heart						
		1996–2000		2000–2003		2003–2008		*Total*	1996–2000		2000–2003		2003–2008		*Total*
From	To	(ha)	(%)	(ha)	(%)	(ha)	(%)	*(%)*	(ha)	(%)	(ha)	(%)	(ha)	(%)	*(%)*
1. Actual urbanisation		*88*	*0.14*	*401*	*0.63*	*191*	*0.30*	*1.1*	*409*	*0.22*	*660*	*0.35*	*1,052*	*0.56*	*1.1*
Agriculture	Built-up	16	0.03	138	0.22	114	0.18		113	0.06	539	0.29	447	0.24	
Nature	Built-up	−8	−0.01	58	0.09	−13	−0 02		−15	−0.01	72	0.04	−3	0.00	
Urban green	Built-up	29	0.05	39	0.06	−56	−0.09		240	0.13	−31	−0.02	254	0.14	
Greenhouses	Built-up	4	0.01	0	0.00	−1	0.00		5	0.00	14	0.01	−3	0.00	
Agriculture	Infrastructure	26	0.04	101	0.16	53	0.08		65	0.03	47	0.02	115	0.06	
Nature	Infrastructure	−1	0.00	−9	−0.01	37	0.06		−26	−0.01	2	0.00	60	0.03	
Urban green	Infrastructure	22	0.03	74	0.12	51	0.09		26	0.01	17	0.01	181	0.10	
2. Potential urbanization		*837*	*1.31*	*193*	*0.30*	*151*	*0.24*	*1.8*	*574*	*0.31*	*772*	*0.41*	*544*	*0.29*	*1.0*
Agriculture	Urban green	645	1.01	145	0.23	103	0.16		499	0.27	625	0.33	467	0.25	
Greenhouses	Urban green	12	0.02	−10	−0.02	−16	−0 02		1	0.00	−24	−0.01	−7	0.00	
Agriculture	Greenhouses	I80	0.28	57	0.09	64	0.10		74	0.04	171	00.9	83	0.04	
3. Nature development		*175*	*0.27*	*168*	*0.26*	*103*	*0.16*	*0.7*	*446*	*0.24*	*98*	*0.05*	*125*	*0.07*	*0.4*
Urban green	Nature	−36	−0 06	3	0.00	13	0.02		20	0.01	−93	−0.05	−20	−0.01	
Agriculture	Nature	235	0.37	142	0.22	100	0.16		417	0.22	203	0.11	166	0.09	
Water	Nature	−24	−0.04	22	0.03	−10	−0.02		10	0.01	−13	−0.01	−22	−0.01	
4. Other changes		*133*	*0.22*	*83*	*0.13*	*40*	*0.06*	*0.4*	*104*	*0.06*	*65*	*0.03*	*136*	*0.07*	*0.2*
Net change (× 1,000 ha; %)		*1.2*	*1.94*	*0.8*	*1.32*	*0.5*	*0.76*	*4.0*	*1.5*	*0.82*	*1.6*	*0.85*	*1.9*	*0.99*	*2.7*
Total surface (× 1,000 ha)		*64*							*187*						

Table 2 Most important net land-use transitions for the National Landscapes (only within the Randstad and excluding the Buffer zones and the Green Heart) and the non-restricted part of the Randstad

Land use change (Net)		National Landscapes (excl. other restricted zones)							Randstad excluding restricted zones						
		1996–2000		2000–2003		2003–2008		Total	1996–2000		2000–2003		2003–2008		*Total*
From	To	(ha)	(%)	(ha)	(%)	(ha)	(%)	(%)	(ha)	(%)	(ha)	(%)	(ha)	(%)	*(%)*
1. Actual urbanisation		*121*	*0.15*	*307*	*0.39*	*326*	*0.41*	*1.0*	*2,447*	*0.87*	*2,243*	*0.79*	*4,157*	*1.47*	*3.1*
Agriculture	Built-up	101	0.13	106	0.13	131	0.17		943	0.33	423	0.15	1,063	0.38	
Nature	Built-up	−6	−0.01	−3	0.00	16	0.02		−16	−0.01	50	0.02	134	0.05	
Urban green	Built-up	52	0.07	55	0.07	78	0.10		1,401	0.50	1,385	0.49	2,275	0.81	
Greenhouses	Built-up	0	0.00	1	0.00	3	0.00		82	0.03	46	0.02	116	0.04	
Agriculture	Infrastructure	−17	−0 02	117	0.15	46	0.06		38	0.01	160	0.06	216	0.08	
Nature	Infrastructure	−2	0.00	8	0.01	10	0.01		−41	−0.01	−12	0.00	30	0.01	
Urban green	Infrastructure	−7	−0.01	23	0.03	43	0.05		39	0.01	186	0.07	322	0.11	
2. Potential urbanisation		*401*	*0.51*	*44*	*0.06*	*343*	*0.43*	*1.0*	*4,613*	*1.63*	*2,735*	*0.97*	*2,452*	*0.87*	*3.5*
Agriculture	Urban green	376	0.47	21	0.03	320	0.40		4,058	1.44	2,378	0.84	2,202	0.78	
Greenhouses	Urban green	2	0.00	0	0.00	13	0.02		206	0.07	329	0.12	311	0.11	
Agriculture	Green houses	24	0.03	23	0.03	11	0.01		348	0.12	29	0.01	−61	−0.02	
3. Nature development		*131*	*0.17*	*81*	*0.10*	*70*	*0.09*	*0.4*	*127*	*0.05*	*489*	*0.17*	*298*	*0.11*	*0.3*
Urban green	Nature	−4	−0.01	13	0.02	1	0.00		−141	−0.05	265	0.09	190	0.07	
Agriculture	Nature	132	0.17	74	0.09	73	0.09		281	0.10	227	0.08	87	0.03	
Water	Nature	3	0.00	−7	−0.01	−4	0.00		−14	0.00	−2	0.00	21	0.01	
4. Other changes		*44*	*0.06*	*73*	*0.09*	*94*	*0.12*	*0.3*	*170*	*0.06*	*219*	*0.08*	*437*	*0.15*	*0.3*
Net change (× 1,000 ha;%)		*0.7*	*0.88*	*0.5*	*0.64*	*0.8*	*1.05*	*2.6*	*7.4*	*2.61*	*5.7*	*2.01*	*7.3*	*2.60*	*7.2*
Total surface (× 1,000 ha)		*79*							*282*						

We have chosen to distinguish urbanisation in two processes: actual and potential urbanisation. Actual urbanisation refers to a change into built-up area or infrastructure, whereas potential urbanisation refers to the new development of urban green or greenhouses. The latter process indicates the development of land uses with a distinct urban appearance that, especially in the case of urban green, often precede various types of built-up land (such as residences or businesses). Yet, these types of land use are not yet fully urbanised and transitions back to more rural types of land use are still possible. This reverse process is much more unlikely in the case of built-up land. To highlight the process of nature development that is supposed to be especially prevalent in the Buffer zones we have also summarised transitions related that indicate new nature areas. The remaining minor transitions are summarized under 'Other changes'. These include all other possible transitions that are not listed under the three main processes listed above, such as those from: water to urban area, water to greenhouses, greenhouses to Infrastructure et cetera. To be able to compare the amount of transition between different zones a percent change per transition is calculated by dividing the amount of change by the total surface of the respective zone.

The tables only show net transitions because we want to focus on the dominant spatial processes. So when we present the transitions from, for example, Agriculture to Built-up, this is the result of all transitions from Agriculture to Built-up minus the transitions from Built-up to Agriculture in that period in the respective region. Pixel-to-pixel comparisons of these highly detailed data sets highlight many small-scale transitions that are partly related to mapping and classification issues such as changes in the minimum mapping unit from year to year. These issues are acknowledged by the data provider (Melser 2012). The high temporal resolution and distinction of three subsequent short transitions periods also increases the amount of observed change: inner-city redevelopment processes, for example, may be observed as an urban area to urban green transition in period 1 and an urban green to urban area transition in period 2. Similarly the development of agricultural land into natural areas (woodlands) may follow an intermediate urban green (vacant land) stage. The combination of classification and mapping issues and short-term temporal changes makes that our three-period description of transitions show 35 % more change than a one-step assessment of transitions between 1996 and 2008 does. We have chosen for this high temporal resolution, however, to be able to study potential changes in urbanisation speed that may be the result of policy changes over time.

The tables show that the non-restricted part of the Randstad is by far the most dynamic area in the Randstad with a total net change of about 20,000 ha in the total observed 12-year period. This is equivalent to about 7 % of the total surface area of the non-restricted part of the Randstad and indicates that approximately 0.6 % of this region changes its use every year. The other areas are much less dynamic: the total net change in the Buffer zones is equivalent to about 4 % of that area, whereas this share is about 3 % in both the Green Heart and the additional area that has more recently been designated a National Landscape within the Randstad. These dynamics mainly refer to actual and potential urbanisation processes that typically are equally strong in each region, claiming in total about 3.1 % and 3.5 % respectively of the total non-restricted region in the Randstad and about 1 % each in the other regions in the total studied period. When we focus on the restricted landscapes we find that in the Buffer zones both potential urbanisation and nature development are more prevalent (claiming 1.8 % and 0.7 % respectively of the total Buffer zone area) than in the Green Heart and National Landscapes (claiming 1.0 % and 0.4 % respectively of these areas). This may point at the strong policy attention to the Buffer zone where often recreational facilities and natural areas were being developed (van Rij et al. 2008).

When we look in more detail at the process of actual urbanisation we find that in most regions the land that becomes built-up was previously in agricultural use. Only within the non-restricted areas of the Randstad this used to be mainly urban green, probably indicating that in these areas agricultural land is relatively scarce and green forms of urban land (allotment gardens, small green spaces and building lots) are more prevalent and apparently also likely to become urbanised. Nature areas hardly change into urban types of land use. They are probably better able to withstand the urbanisation pressure because of specific (inter)national nature and biodiversity protection policies and the reluctance of nature land owners to sell their land for urbanisation. Natural land in the Netherland is typically owned by Governmental or not-for-profit nature conservation organisations that have the specific objective of preserving natural areas.

The temporal dynamics do not offer a clear signal; the relative strength of the observed processes varies over time in a different way for each region. These dynamics do not seem to be related with policy changes over time and may, in fact hint at the complex and coincidental relation between planning, economic opportunities and spatial developments. Exactly which development will take place at a particular location at a certain moment in time will depend on many factors. The Dutch planning system is notoriously slow in keeping up with changes (Van Rij 2009); spatial plans may need a long time to become reality because of the various governmental layers that may be involved and (legal) procedures that have to be followed. For the same reasons spatial restrictions may linger on for quite some time after they have been abolished at one governmental layer. Once development plans are approved and implemented it may still take several years before the final results can be observed because of lengthy construction processes (in particular with infrastructure development) or altered socio-economic conditions (e.g. changing societal consumer preferences, sudden financial crises). So developments that are observed at a particular year are not necessarily related to the planning and socio-economic conditions of that moment.

3.2 *Object-Based Analysis*

In addition to the raster-based analysis that focussed on land-use changes we also analysed regional changes in the total number of houses. This object-based analysis of residential development focuses on two periods of three respectively 4 years between 2000 and 2008. Table 3 shows the annual growth in housing stock, housing densities and urbanisation rates within different (restricted and non-restricted) parts of the Randstad and for the Netherlands as a whole.

The results show that housing density is about five times higher in the Randstad (excluding protected zones) than the Dutch average. In absolute terms, the annual growth in number of houses in both periods is by far the largest in this zone, further enlarging the existing differences in absolute numbers and housing density. However, observing growth in relative terms, we find a substantial increase within the restricted zones, partly because of the low initial number of houses. In combination with low amounts of additional urban land within these restricted zones (recall our preceding analysis in this chapter) this indicates that a substantial part of new houses in these zones is realised within existing urban areas. Unfortunately we cannot analyse that in more detail as the spatial detail of this data is insufficient (or rather, not reliable enough).

Compared to the national totals, the growth of the housing stock within the Randstad is relatively low. This implies that other parts of the country accommodate a larger share of the new houses. This happens mainly at New towns and designated urban extension locations in the provinces that neighbour the Randstad region (most notably Flevoland and Noord-Brabant). This development may indicate that the Randstad is reaching its limits for further urban development in the currently common form that is dominated by a large share of single-family dwellings.

From Table 3 we can also observe that housing density is by far the lowest in the Buffer zones (87 houses per km^2 in 2008, even lower than in the rural, peripheral zone of the country). This shows that these zones are still truly open spaces where urban development is successfully limited.

3.3 Assessment of Potential Future Changes

The preceding sections have indicated that restrictive spatial policy is (partially) successful in limiting urban development in specific regions. This knowledge can be used to simulate potential future conditions. As part of the Strategic Environmental Assessment of the proposed changes in national spatial policy discussed extensively in Sect. 1.3, we applied Land Use Scanner to simulate the possible future urbanisation patterns that might arise under different policy scenarios. This section describes the way these simulations were constructed. For more information about the policy context and expected environmental impacts the reader is referred to the actual assessment report (Elings et al. 2011). The model uses a spatially aggregated version of the 2008 land-use raster data set from Statistics Netherlands as starting point for simulation at a 100 m resolution.

The new national spatial strategy proposes four sets of major changes from current national spatial policy:

Table 3 Regional increases in number of houses and housing densities in different zones within the Randstad (2000–2008)

Region	Total surface area (km^2)	2000 (abs.)	2000 ($/km^2$)	Annual change 2000–2003 (abs.)	Annual change 2000–2003 (%)	Annual Change 2004–2008 (abs.)	Annual Change 2004–2008 (%)	2008 (abs.)	2008 ($/km^2$)
Randstad*	2,822	2,542,943	901	11,808	0 46	14,780	0.58	2,614,355	926
Buffer zones	639	52,084	82	368	0.71	270	0.50	55,302	87
Green Heart	1,872	279,638	149	2,406	0.86	1,941	0.67	297,785	159
National Landscapes*	791	107,223	136	968	0.90	817	0.74	113,537	144
The Netherlands	*34,993*	*6,843,865*	*196*	*39,306*	*0.57*	*59,678*	*0.85*	*7,248,395*	*207*

*Excluding the other regions (such as Buffer zones) that partially fall within them

1. A new evaluation framework for investments in infrastructure that emphasises economic benefits;
2. Less national interference in urban development, abolishing transformation and urban concentration policy;
3. Limiting the ambitions for the National Ecological Network, focus on management of current natural areas, limited acquisition of new areas, no development of connection zones;
4. Stronger emphasis on internationally unique cultural historic landscape values of, for example, UNESCO world heritage sites, abolishing buffer zones and decentralising national landscapes.

The outcome of these policy changes, however, is uncertain. This is especially true for the impact of decentralising the responsibility for the National Landscapes. Provinces may decide to ease, continue or reinforce the current restrictive planning regime in these areas. To show the uncertainty related to the different possible attitudes of the three provinces involved, it was decided to show two potential, extreme outcomes: one in which current policies are fully maintained (reference alternative) and one in which they are abolished (new policy alternative). Neither outcome is necessarily more likely, but together they show the potential bandwidth of impacts. This scenario-based approach is advocated in strategic planning (Dammers 2000) and decision making (De Ruijter et al. 2011), but not very common in environmental assessment reports as was previously also recognised by Duinker and Greig (2007).

Table 4 summarises the main policy objectives and associated extreme spatial implications of the reference situation and the new policy alternative grouped per policy domain. These alternative-specific assumptions were translated in model input and first fed into the TIGRIS-XL land-use transport interaction model (Zondag and Geurs 2011) to obtain separate sets of regional projections of the demand for new residences and business estates for each alternative in 2040. Figure 5 presents an overview of the demand for urban land according to the Current and New policy alternatives compared to the historic trend. The figure makes clear that the additional demand for urban land for both alternatives is almost identical. These projected developments are, furthermore, in line with the trend in the past 30 years.

In a subsequent step these regional demands for additional urban land, together with alternative-specific, spatially explicit assumptions related to, for example, the presence (or absence) of specific policy restrictions, were fed into Land Use Scanner to simulate land-use patterns. These simulations were carried out by PBL Netherlands Environmental Assessment Agency as part of their ex-ante evaluation of the new policy report (Kuiper and Evers 2011a) and build upon initial work that was done for the ‘Netherlands in the future’ study (PBL 2010; Kuiper et al. 2011).

Figure 6 shows the simulated increase in urban area for the reference alternative that follows current policy and the new policy alternative. The figure essentially shows three states: new urban areas according to current policy, new urban areas according to the new policy alternative and new urban areas according to both

Table 4 Overview of the main objectives and expected spatial implications of the Current policy (reference situation) and New policy alternatives grouped per policy domain

Policy domain	Current policy	New policy alternative
1. Mobility and accessibility	Current plans for development new infrastructure carried out	Stronger focus on economic benefits, but spatial implications uncertain: current plans are maintained
2. Urbanisation (residence/ commerce)	Bundling and transformation zones maintained, results in ca 30 % intensification (share of new residences built in current urban areas)	Bundling and transformation zones abolished: ca. 20 % intensification
	Supply steers location of new residences	Residential preferences and accessibility (demand) dominate location of new residences
3. Nature development	National Ecological Network realised in 2018 according to initial plan: 100,000 ha nature extra	Limited version of National Ecological Network: 20,000 ha extra
4. Unique landscape values	Buffer zones and National Landscapes limit urbanisation in specific areas	Buffer zones abolished, limited impact National Landscapes
	National Ecological Network and Natura 2000 areas limit urbanisation	Only international obligations limit urbanisation (UNESCO, Natura 2000)

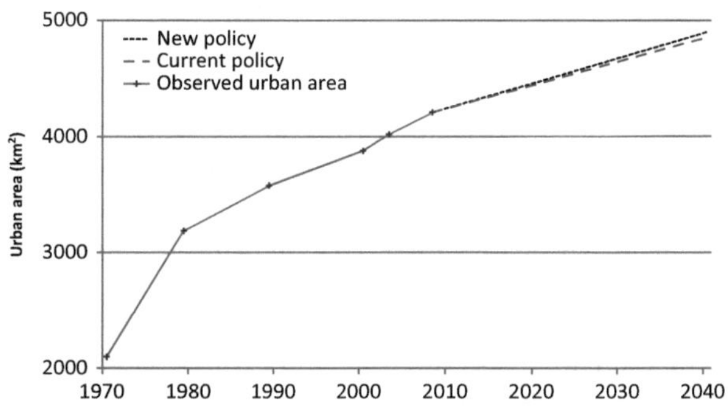

Fig. 5 Observed and projected urban area according to the Current and New policy alternatives (Source observed urban area: Statistics Netherlands 2012)

alternatives. The latter thus show locations that are likely to become urbanised irrespective of any changes in spatial policy. To visualise the uncertainty that is inherent to the simulation outcomes, the very detailed outcomes are not directly shown at their initial 100 m resolution. Instead, a visualisation technique is chosen that emphasises the presence of similar neighbours in a 500 m environment. With a moving window filter the 24 cells surrounding a central cell are evaluated; when all cells show the same value as the central cell a colour with a high intensity is

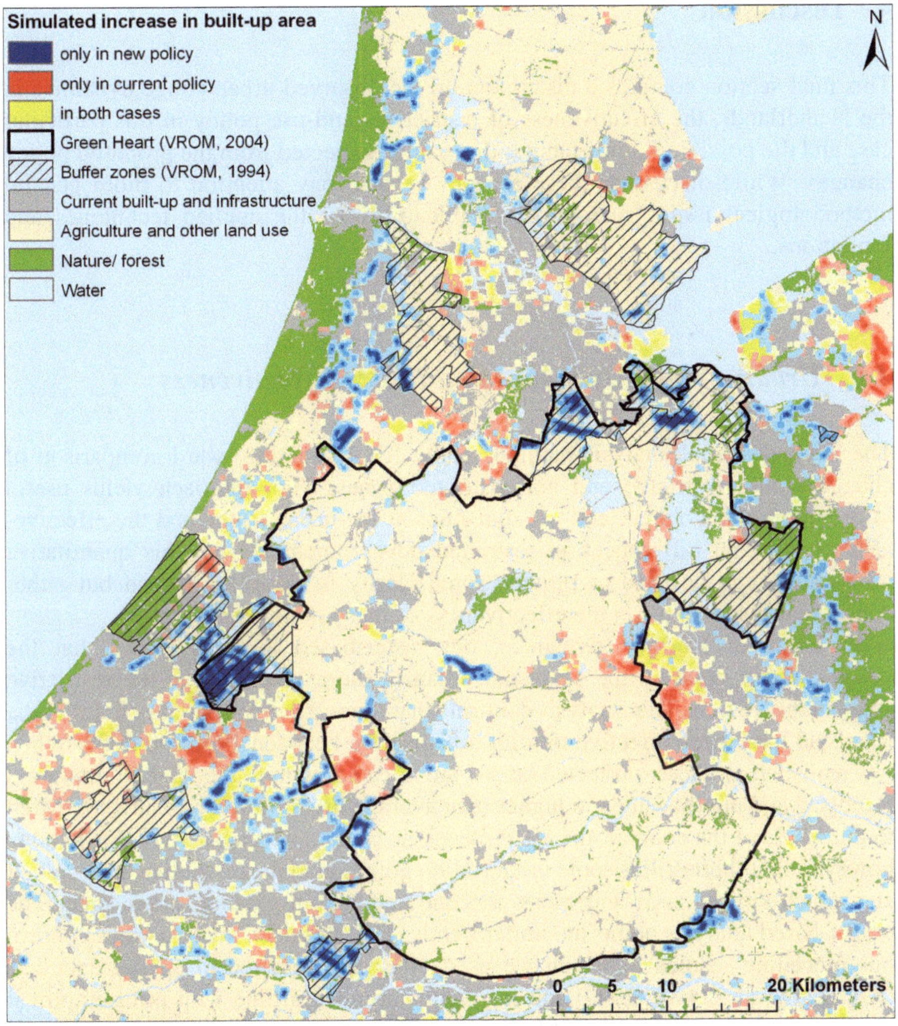

Fig. 6 Simulated changes in built-up area for the new national spatial strategy and current policy (adapted from Kuiper and Evers 2011a). Colour intensity in simulated built-up area represents the amount of similar neighbours, so intensely coloured locations indicate likely large-scale urban development under specific policy conditions

selected, when no other neighbour has the same value a low intensity colour is selected.

Obviously the simulations offer indicative, almost caricatural images that do not allow detailed impact assessments with environmental impact models. But these simulations integrate the potential implications of domain-specific policies and help visualising the regional accumulation of the impacts associated with individual policy measures.

4 Discussion

This final section contains a discussion on the observed urban fringe dynamics in the Netherlands, the effectiveness of restrictive land-use policy in this particular case and the possible future changes that can be expected from the proposed policy changes. While discussing these aspects we also pay attention to more general methodological issues such as the applicability of the applied techniques and limitations.

4.1 Urban Fringe Dynamics and Policy Effectiveness

The applied spatial analysis methods allow for a straightforward comparison of change within restricted and non-restricted zones. This approach yields useful information about actual land-use dynamics at the urban fringe and the effectiveness of regional differences in restricted development zones. This quantitative information can be used to provide some clarity in the often heated but rather conceptual discussions on the effectiveness of these policy measures

In terms of land-use dynamics, our raster-based analysis shows that the non-restrictive areas in the Randstad urbanise more quickly than the restrictive areas in this region. This analysis confirms that the restrictive spatial policies in the Randstad have been effective to limit urbanisation between 1996 and 2008. When we realise that these protected areas are located within the highly urbanised Randstad and are thus under a higher than average urbanisation pressure compared to the whole of the country, the effectiveness is even more impressive. We do note, however, that there still is some urbanisation going on in the restrictive areas, albeit at a much slower pace and more concentrated within existing urban area. The object-based analysis points at similar trends. The presented analyses thus confirm our preceding studies in which we used other and less recent spatial data sets (Koomen et al. 2008b, 2009) and the results are also in line with prior empirical work of others (VROM 2000; MNP 2004). The importance of zoning regulations was also acknowledged in an in-depth case study performed by Van Rij et al. (2008) that relied on literature reviews and interviews. They concluded that zoning regulations together with the development of local recreational potential were indeed effective in limiting urbanisation within the restricted zones. The developments that are taking place in the restrictive areas, for instance in the Green Heart, are urban extensions in a limited number of municipalities that are clearly in line with the compact-city philosophy that had a central role in Dutch spatial planning for the past decades.

In general, in both our analyses we do not observe a clear difference between the different restrictive areas, or between different periods. Thus we cannot observe impacts of, for example, the policy-shift in 2004. This might be partially attributed to the fact that the study period is relatively short and the policy change establishing the National Landscapes is still rather recent. It may, in fact, show that planning is a slow and complex process, the outcome of which relates to, amongst others, a degree of inertia in the planning system, lengthy construction processes and changing societal, planning and economic conditions.

It is important to stress that the relative success of open space preservation policies was also due to specific conditions and additional policies that were in place (Dieleman et al. 1999). To keep specific areas open, policy makers in the past decades felt the need to improve their recreational, natural and agricultural potential. In addition to the special zoning status, land consolidation and, from 1964 onwards, also land acquisition were seen as appropriate instruments (Bervaes et al. 2001). It is, furthermore, important to note that the Buffer zone policy has been drafted together with specific plans (growth centres and growth towns) to steer urbanisation towards the outer edges of the Randstad (Faludi and Van der Valk 1994). The restrictions on urbanisation inside the Randstad were thus compensated with urbanisation incentives outside of it.

Obviously, the presented results do not provide information about the societal, economic or other forces that drive the observed changes. Apart from zoning regulations, other factors such as more limited accessibility, employment and service levels, may have influenced the regional differences in urban development. These factors are not considered in this chapter but will be considered for future explanatory studies that will apply regression analysis to link various spatially explicit driving forces (including spatial restrictions) with observed (changes in) urbanisation patterns.

Our analyses were hampered by limited data availability and data definition issues, leading to relatively short periods of analysis and some inconsistencies in the data. Fortunately, by combining two analyses of different data sets we were able to obtain a fairly robust idea about the relevant land-use dynamics in the area over the past decade. The use of local-level geographical data also allowed us to graphically represent land-use changes and visually analyse the patterns of change. That is a clear advantage over comparable efforts that rely on, for example, census statistics (see, for instance, Kline 2000; Nelson 2004). The object-based data did not allow detailed local assessments of urban development, but was found to provide reliable statistics at the regional level that correspond with other sources such as Statistics Netherlands.

4.2 *Potential Impacts of Proposed Policy Changes*

In addition to our GIS-based analyses of past land-use changes we applied simulation methods to assess potential impact of proposed policy changes. This approach provided a valuable tool to depict likely outcomes of policy changes and was successfully used in a recent Strategic Environmental Assessment report dedicated to the newly proposed national spatial strategy for the Netherlands. This what-if type of simulation typically relies strongly on expert judgement as it has to describe the impact of policies that have not yet been implemented and whose effects cannot be observed. The GIS-based analyses of the effectiveness of similar policies as described in this chapter thus offer an important ingredient for the development and calibration of land-use simulation models.

A recent, interview-based study by Van Kouwen (2012) into the potential future spatial developments in National Landscapes made clear that most parties involved do not expect sudden changes from the proposed changes in national spatial policy. Representatives of governmental organisations at the national, regional and local level and societal partners base these expectations on their observations that: (1) provinces have already implemented current, more restrictive policies in their strategic visions that govern planning at the municipal level; and (2) new urban developments are unlikely under current economic conditions. These expectations, however, seem to be strongly linked to current political and economic conditions. When new regional governments are formed and the economy recovers, the current trend towards deregulation and decentralisation may lead to an increase in urban development in currently protected landscapes. Such threats to metropolitan open spaces have, for example, also been described by Van Rij (2009). Therefore, we strongly believe that the type of scenario-based simulations of potential impacts of policy changes described in this chapter offer a powerful approach to incorporate the notion of uncertainty in ex-ante policy evaluation.

Acknowledgements This research is funded by the research programmes Sustainable earth of the Netherlands organisation for scientific research (NWO) and Knowledge for Climate. We would like to thank PBL Netherlands Environmental Assessment Agency for providing the spatial data and land-use simulations that were incorporated in the analyses discussed in this chapter. Part of the work described here was performed as part of a project one of the authors carried out for the Ministry of Infrastructure and Environment in cooperation with Royal Haskoning and Alterra Wageningen University and Research Centre. The former colleagues and partners in that project are thanked for the pleasant and effective cooperation and fruitful discussions that provided the results discussed here. Finally, we also want to thank our colleague Ronnie Lassche for his reflections on the methodologies of our analyses.

References

Alpkokin, P. (2012). Historical and critical review of spatial and transport planning in the Netherlands. *Land Use Policy, 29*(3), 536–547.

Alterman, R. (1997). The challenge of farmland preservation: Lessons from a six-nation comparison. *Journal of the American Planning Association, 63*(2), 220–243.

Bae, C.-H. C., & Richardson, H. W. (2004). Introduction. Chapter 1. In H. W. Richardson & C.-H.C. Bae (Eds.), *Urban sprawl in Western Europe and the United States* (pp. 1–7). Aldershot: Ashgate.

Bartlett, J. P., Mageen, D. M., & O'Conner, R. J. (2000). Residential expansion as a continental threat to US coastal ecosystems. *Population and Environment, 21*(5), 429–469.

Bervaes, J. C. A. M., Kuidersma, W., & Onderstal, J. (2001). *Rijksbufferzones; verleden, heden en toekomst*. Alterra rapport 360. Wageningen: Alterra.

Borsboom-van Beurden, J. A. M., Bakema, A., & Tijbosch, H. (2007). A land-use modelling system for environmental impact assessment; recent applications of the LUMOS toolbox. Chapter 16. In E. Koomen, J. Stillwell, A. Bakema, & H. J. Scholten (Eds.), *Modelling land-use change; progress and applications* (pp. 281–296). Dordrecht: Springer.

Burke, G. L. (1966). *Greenheart metrolopolis*. London: Macmillan.

CBS. (2008). *Bestand bodemgebruik productbeschrijving*. Voorburg/Heerlen: Centraal Bureau voor de Statistiek.

Dammers, E. (2000). *Leren van de toekomst. Over de rol van scenario's bij strategische beleidsvorming*. Delft: Uitgeverij Eburon.

De Ruijter, P., Stolk, S., & Alkema, H. (2011). *Klaar om te wenden; handboek voor de strateeg*. Schiedam: Scriptum.

Dekkers, J. E. C., & Koomen, E. (2007). Land-use simulation for water management: Application of the land use scanner model in two large-scale scenario-studies. Chapter 20. In E. Koomen, J. Stillwell, A. Bakema, & H. J. Scholten (Eds.), *Modelling land-use change; progress and applications* (pp. 355–373). Dordrecht: Springer.

Dieleman, F. M., Dijst, M. J., & Spit, T. (1999). Planning the compact city: The Randstad Holland experience. *European Planning Studies, 7*(5), 605–621.

Duinker, P. N., & Greig, L. A. (2007). Scenario analysis in environmental impact assessment: Improving explorations of the future. *Environmental Impact Assessment Review, 27*(3), 206–219.

EL&I. (2009). *Verburg en Cramer selecteren 10 voorbeeldprojecten aanpak verspreid liggende kassen*. The Hague: Ministerie van Economische Zaken, Landbouw en Innovatie.

Elings, C., Zijlstra, R., Koomen, E., & De Groot, S. (2011). *Milieueffectrapport Ontwerp Structuurvisie Infrastructuur en Ruimte in opdracht van Ministerie van Infrastructuur en Milieu*. Amsterdam/Nijmegen: Geodan/Royal Haskoning.

Evers, W., Vries, L., De Man, R., & Schotten, C. G. J. (2005). *Woning- en populatiebestanden in het dataportaal. Overzicht van de basisbestanden, bewerkingen en kwaliteitsacties*. Bilthoven: RIVM.

Faludi, A., & Van der Valk, A. (1994). *Rule and order: Dutch planning doctrine in the twentieth century*. Dordrecht: Kluwer Academic Publishers.

Frenkel, A. (2004). The potential effect of national growth management policy on urban sprawl and the depletion of open spaces and farmland. *Land Use Policy, 21*(4), 357–369.

Gailing, L. (2005). *Regionalparks, Grundlagen und Instrumente der Freiraumpolitik in Verdichtungsräume, Blaue Reihe, Dortmunder Beiträge zur Raumplanung 121*. Dortmund: Institut für Raumplanung Universität Dortmund.

Hall, P. (1966). *The world cities*. London: World University Press.

I&M. (2011). *Ontwerp structuurvisie infrastructuur en ruimte; Nederland concurrerend, bereikbaar, leefbaar en veilig*. Den Haag: Ministerie van Infrastructuur en Milieu.

Jacobs, C. G. W., Koomen, E., Bouwman, A. A., & Van der Burg, A. (2011). Lessons learned from land-use simulation in regional planning applications. Chapter 8. In E. Koomen & J. Borsboom-van Beurden (Eds.), *Land-use modelling in planning practice* (pp. 131–149). Heidelberg: Springer.

Jongman, R. H. G., Knlvik, M., & Kristiansen, I. (2004). European ecological networks and greenways. *Landscape and Urban Planning, 68*(2–3), 305–319.

Kline, J. (2000). Comparing states with and without growth management, analysis based on indicators with policy indications comment. *Land Use Policy, 4*(17), 349–355.

Klosterman, R. E. (1999). The what if? Collaborative planning support system. *Environment and Planning B: Planning and Design, 26*, 393–408.

Koomen, E., & Borsboom-van Beurden, J. (Eds.). (2011). *Land-use modeling in planning practice*. Geojournal library, vol. 101. Heidelberg: Springer.

Koomen, E., Rietveld, P., & De Nijs, T. (2008a). Modelling land-use change for spatial planning support; editorial. *Annals of Regional Science, 42*(1), 1–10.

Koomen, E., Dekkers, J., & Van Dijk, T. (2008b). Open-space preservation in the Netherlands: Planning, practice and prospects. *Land Use Policy, 25*(3), 361–377.

Koomen, E., Geoghegan, J., & Dekkers, J. E. C. (2009). The impact of open space preservation policies; evidence from the Netherlands and the US. Chapter 9. In A. van der Valk, T. Van Dijk (Eds.), *Regional planning for open space* (pp. 179–198). London: Routledge.

Koomen, E., Hilferink, M., & Borsboom-van Beurden, J. (2011a). Introducing land use scanner. Chapter 1. In E. Koomen & J. Borsboom-van Beurden (Eds.), *Land-use modeling in planning practice* (pp. 3–21). Dordrecht: Springer.

Koomen, E., Koekoek, A., & Dijk, E. (2011b). Simulating land-use change in a regional planning context. *Applied Spatial Analysis and Policy, 4*(4), 223–247.

Korthals Altes, W. (2009). Taxing land for urban containment: Reflections on a Dutch debate. *Land Use Policy, 26*(2), 233–241.

Kuiper, R., & Evers, D. (2011a). *Ex-ante evaluatie Structuurvisie Infrastructuur en Ruimte.* Den Haag: Planbureau voor de Leefomgeving.

Kuiper, R., & Evers, D. (2011b). Structuurvisie Infrastructuur en Ruimte perkt nationale belangen in. *ROM Magazine, 29*(6), 10–13.

Kuiper, R., Kuijpers-Linde, M., & Bouwman, A. A. (2011). A sustainable outlook on the future of the Netherlands. Chapter 4. In E. Koomen & J. Borsboom-van Beurden (Eds.), *Land-use modelling in planning practice* (pp. 61–77). Heidelberg: Springer.

Landis, J. D. (1994). The California urban futures model: A new generation of metropolitan simulation models. *Environment and Planning B: Planning and Design, 21*, 399–420.

Longley, P., Batty, M., Shepherd, J., & Sadler, G. (1992). Do green belts change the shape of urban areas? A preliminary analysis of the settlement geography of South East England. *Regional Studies, 26*(5), 437–452.

Loonen, W., & Koomen, E. (2009). *Calibration and validation of the land use scanner allocation algorithms*. PBL-report. Bilthoven: Netherlands Environmental Assessment Agency.

Maruani, T., & Amit-Cohen, I. (2007). Open space planning models: A review of approaches and methods. *Landscape and Urban Planning, 81*, 1–13.

Melser, C. (2012). Het bestand bodemgebruik gaat met de tijd mee. *Geo-Info, 4*, 4–7.

MNP. (2004). *Milieu- en natuureffecten Nota Ruimte*. RIVM-rapport 711931009. Bilthoven: RIVM.

Nelson, A. C. (2004). Urban containment American style: A preliminary assessment. In H. W. Richardson & C.-H. C. Bae (Eds.), *Urban sprawl in Western Europe and the United States*. Aldershot: Ashgate.

Ottens, H. F. L. (1979). Spatial development in the Green Heart of the Randstad: Policies versus theoretical and empirical evidence. *Tijdschrift voor Economische en Sociale Geografie, 70*(3), 130–143.

PBL. (2010). *The Netherlands in the future. Second sustainability outlook: The physical living environment in the Netherlands*. Bilthoven: PBL Netherlands Environmental Assessment Agency.

Pena, J., Bonet, A., Bellot, J., Sanchez, J. R., Eisenhuth, D., Hallett, S., & Aledo, A. (2007). Driving forces of land-use change in a cultural landscape of Spain. Chapter 6. In E. Koomen, J. Stillwell, H. J. Scholten, & A. Bakema (Eds.), *Modelling land-use change; progress and applications* (pp. 97–115). Dordrecht: Springer.

Pieterse, N., Van der Wagt, M., Daalhuizen, F., Piek, M., Künzel, F., & Aykaç, R. (2005). *Het gedeelde land van de Randstad. Ontwikkeling en toekomst van het Groene Hart*. Rotterdam/ Den Haag: NAi Uitgevers/RPB.

Rietveld, P. (2001). Urban development between market forces and policy intervention; experiences in the Netherlands. *Journal of Japan Real Estate Society, 5*, 23–30.

RNP. (1958). *De ontwikkeling van het Westen des Lands*. Den Haag: Rijksdienst voor het Nationale Plan.

Romero, F. S. (2003). Open space preservation policies: An institutional case study. *Journal of Architecture and Planning Research, 20*(2), 146–174.

Roodbol-Mekkes, P. H., Van der Valk, A. J. J., & Korthals Altes, W. K. (2012). The Netherlands spatial planning doctrine in disarray in the 21st century. *Environment and Planning A, 44*(2), 377–395.

Scholten, H. J., Van de Velde, R. J., Rietveld, P., & Hilferink, M. (1999). Spatial information infrastructure for scenario planning: The development of a land use planner for Holland. In J. Stillwell, S. Geertman, & S. Openshaw (Eds.), *Geographical information and planning* (pp. 112–134). Berlin/Heidelberg/New York: Springer.

Schotten, C. G. J., Goetgeluk, R. W., Hilferink, M., Rietveld, P., & Scholten, H. J. (2001). Residential construction, land use and the environment. Simulations for The Netherlands using a GIS-based land use model. *Environmental Modelling and Asessment, 6*(2), 133–143.

Schwanen, T., Dieleman, F. M., & Dijst, M. (2001). Travel behaviour in Dutch monocentric and policentric urban systems. *Journal of Transport Geography, 9*(3), 173–186.

Statistics Netherlands. (2012). Statline, the central database of Statistics Netherlands. http://www.statline.nl. Accessed 1 Aug 2012.

Tesser, P. T. M., Van Praag, C. S., Van Dugteren, F. A., Herwijer, L. J., & Van der Wouden, H. C. (1995). *Rapportage minderheden 1995. Concentratie en segregatie. SCP-cahier 123*. Rijswijk/Den Haag: Sociaal en Cultureel Planbureau/VUGA.

V&B. (1960). *Nota inzake de ruimtelijke ordening in Nederland*. Den Haag: Ministerie van Volkshuisvesting en Bouwnijverheid, staatsdrukkerij en uitgeverijbedrijf.

V&RO. (1966). *Tweede Nota over de Ruimtelijke Ordening*. Den Haag: Ministerie van Volkshuisvesting en Ruimtelijke ordening, Staatsuitgeverij.

V&RO. (1977). *Derde Nota over de Ruimtelijke Ordening (verstedelijkingsnota)*. Den Haag: Ministerie van Volkshuisvesting en Ruimtelijke ordening, Staatsuitgeverij.

Van der Hoeven, E., Aerts, J., Van der Klis, H., & Koomen, E. (2009). An integrated discussion support system for new Dutch flood risk management strategies. Chapter 8. In S. Geertman & J. C. H. Stillwell (Eds.), *Planning support systems: Best practices and new methods* (pp. 159–174). Berlin: Springer.

van der Valk, A. (2002). The Dutch planning experience. *Landscape and Urban Planning, 58* (2–4), 201–210.

van der Valk, A., & Van Dijk, T. (Eds.). (2009). *Regional planning for open space. The RTPI library series*. Abingdon: Routledge.

Van Kouwen, M. (2012). *Nationale Landschappen op de schop. Vernieuwing of achteruitgang? Bachelor thesis earth and economics*. Amsterdam: Vrije Universiteit.

van Leeuwen, N. (2004). Bestand bodemGebruik en Top10Vector geharmoniseerd; CBS vernieuwt Bestand BodemGebruik. *Geo-Info, 5*, 218–222.

Van Rij, E. (2009). Threats to metropolitan open space; The Netherlands economic and institutional dimension. Chapter 3. In A. Van der Valk & T. Van Dijk (Eds.), *Regional planning for open space* (pp. 41–60). Abingdon: Routledge.

Van Rij, E., Dekkers, J., & Koomen, E. (2008). Analysing the success of open space preservation in the Netherlands: The Midden-Delfland case. *Tijdschrift voor Economische en Sociale Geografie, 99*(1), 115–124.

VROM. (1989). *Vierde Nota over de Ruimtelijke Ordening*. Den Haag: Ministerie van Volkshuisvesting, Ruimtelijke ordening en Milieubeheer, SDU uitgeverij.

VROM. (2000). *Balans ruimtelijke kwaliteit*. Den Haag: Ministerie van Volkshuisvesting Ruimtelijke Ordening en Milieubeheer.

VROM, LNV, V&W, & EZ (2004). *Nota Ruimte. Ruimte voor ontwikkeling. Ministeries van Volkshuisvesting, Ruimtelijke Ordening en Milieubeheer, Landbouw, Natuur en Voedselkwaliteit, Verkeer en Waterstaat en Economische zaken*. Den Haag: SDU uitgeverij.

Zhao, P. (2011). Managing urban growth in a transforming China: Evidence from Beijing. *Land Use Policy, 28*(1), 96–109.

Zondag, B., & Geurs, K. (2011). Coupling a detailed land-use model and a land-use and transport interaction model. Chapter 5. In E. Koomen & J. Borsboom-van Beurden (Eds.), *Land-use modeling in planning practice* (pp. 79–95). Dordrecht: Springer.

Long-Term Changes in the Configuration of Agriculture and Natural Areas Around Cities in the Netherlands (1900–1990)

Erez Hatna and Martha M. Bakker

Abstract Cities' influence on the spatial configuration of land in their proximity has presumably changed during the last century as agriculture, cities, and transportation evolved. Investigation of these changes has been limited due to the limited availability of historical maps in digital form. In this chapter, we employ a set of digitized historical land-cover maps in order to compare the spatial distribution of cropland, pasture, and nature surrounding cities in the Netherlands for three time periods: 1900, 1960 and 1990. Our findings suggest that the land cover around cities was relatively stable during these time periods. However, we discovered that, near the perimeter of cities, in 1900, we could discern a clear trend of higher fractions of cropland. These tendencies weakened by the middle of the century and almost completely ceased by 1990.

1 Introduction

According to classic spatial economic theory, distance to agricultural markets – often cities – is an important determinant of the land use pattern. Commodity prices combined with transport costs and perishability of agricultural produce determined the optimal location of different land uses around the circumference of a city (von Thünen 1826). Other yield-determining factors such as biophysical landscape

E. Hatna (✉)
Department of Emergency Medicine, Center for Advanced Modeling in the Social, Behavioral and Health Sciences (CAM), Johns Hopkins University, 5801 Smith Avenue, Davis Building – Suite 3220, Baltimore, MD 21209, USA
e-mail: erezh51@gmail.com

M.M. Bakker
Land Use Planning Group, Department of Environmental Science, Wageningen University, P.O. Box 47, 6700 AA Wageningen, The Netherlands
e-mail: martha.bakker@wur.nl

D. Malkinson et al. (eds.), *Modeling of Land-Use and Ecological Dynamics*, Cities and Nature, DOI 10.1007/978-3-642-40199-2_3, © Springer-Verlag Berlin Heidelberg 2013

properties determined the remaining spatial variability in land utilization (Grigg 1984).

The relationship between cities and their hinterland has presumably changed over time. Cities have evolved, become larger, more numerous and less compact, therefore, leading to more areas being nearer to cities. With the emergence of international agricultural markets, cities are often no longer the trade centres for agricultural produce. Furthermore, transport costs have dramatically decreased while refrigeration and conservation techniques have significantly improved, reducing the need to produce perishable products in areas adjacent or close to the cities. Last, the relative profitability of agricultural sectors changed considerably: while cropland was once more profitable than grassland, this has changed as people started to consume more dairy products and meat (Bakker et al. 2011). To some extent, the same applies for nature areas. With increasing welfare, recreation in nature areas may have become a more viable income source than the production of staple crops.

With so many processes interacting simultaneously, modelling the changing relationships between cities and the land use around them is difficult (White and Engelen 2000). Many processes have opposite effects, and small changes in model parameters may cause a qualitative change in model outcomes (Verburg and Overmars 2009). Empirical studies are often hindered by the lack of available data, In order to obtain a robust assessment of relationships between cities and their hinterland, wide-scale data are needed (Angel et al. 2012). Concurrently, the data needs to be digitized in order to perform GIS operations and statistical analyses (Alo and Pontius 2008). Last, because land use is generally inert, changes occur slowly and long periods of time are required to detect significant changes (Rhind and Hudson 1980). These three requirements (temporal extent, spatial extent, and digital format) limit the availability of suitable data.

Our analysis is based on unique datasets that fulfilled these requirements: digital land cover data of 1900, 1960 and 1990 of the Netherlands. These land cover maps were obtained by digitizing historical topographic maps (Alterra 2012). They provide us the opportunity to study the manner in which land cover composition around urban areas changed during the past century in the perspective of a typical modernized country. In the remainder of this chapter, we describe how we processed the data to obtain the composition of three land cover categories (grassland, cropland, and nature) located at varying distances from a selection of urban agglomerations.

2 Setting

The Netherlands is a Western European country bordered by Germany, Belgium, and the North Sea. As with other developed countries, it experienced significant industrialization and urbanization during the twentieth century. The population increased from approximately five million in 1900, to 11.5 million in 1960, to

15 million in 1990 (CBS 2012). Meanwhile, the proportion of people residing in urban areas increased even more rapidly (quantitative estimates vary with the definition of cities (Stone 1975)).

Meanwhile, agriculture was strongly influenced by technological developments. The invention and distribution of artificial fertilizer resulted in large-scale cultivation of heathland. Land reclamations occurred, of which the Flevo Polders are the most spectacular example. Less spectacular, but just as important, were the small-scale land reclamations that converted wetlands, swamps and lakes into fertile agricultural land. Following World War II, large-scale land consolidations occurred, allowing for increased efficiency in land use.

Subsequently, the Common Agricultural Policy of the European Union strongly impacted land use in the Netherlands. The average farm size increased significantly, and most farms specialized in one particular product. Currently, dairy farming is a prominent sector in both economic and land use terms. Arable farming has become firmly dependent on EU subsidies, inherently for sugar beets. Horticulture under glass is significant in economic terms but less so in terms of land use (less than 1 % of the overall agricultural area). Other types of agriculture that are economically important, but not in terms of land use, are various forms of intensive livestock breeding (mostly pigs and poultry).

Despite being a traditionally agricultural country, the Netherlands possesses considerable natural areas, of which the Veluwe, in the centre of the country, is the largest (also one of the largest forests in Europe). Partly these areas were exempted from agriculture because of unfavourable biophysical conditions (poor sandy soils), but they were also privately owned, hence protected, for recreation purposes. In the course of the twentieth century, nature organizations (both private and governmental) purchased these nature areas for both conservation and recreational purposes.

Rather typical for the Netherlands is the strong degree of spatial planning. Together with the United Kingdom, the Netherlands is often seen as an example of quality zoning policy (Van der Valk and De Vries 1996). Zoning plans were previously designed by the central government and have always been strictly imposed. This has reduced urban sprawl; at the same time, it also designated zones for nature development mostly at the expense of agricultural land. Although the spatial planning policy has recently become more decentralized, and became less efficient as a result, during the period studied in this chapter, efficient top-down zoning policy was still in place.

3 Data

Three land cover raster maps were employed to extract land-cover data. These maps include the HGN1900, HGN1960 and HGN1990 maps from the Alterra Institute of Wageningen University (Altera 2012). The three maps encapsulate the entire Netherlands: the HGN1900 map has a cell size of 50 m while the HGN1960 and

Table 1 The land cover aggregation used in the study

Original category	Newly derived category
Pasture	Pasture
Cropland	Cropland
Built-up area	Built-up area
Greenhouses	Greenhouses
Water	Water
Heathland	Nature
Deciduous forest	
Confiner forest	
Reed swamp	
Sands and dunes	

HGN1990 maps have a cell size of 25 m. The HGN1900 map was constructed from analogue topographical maps and has an accuracy of 96 % (Knol et al. 2004). The HGN1960 and HGN1990 were resampled using a majority rule into 50 m cell size to create compatibility with the 1900 map.

The original HGN1900 map contains 10 classes (Table 1); the HGN1960 and HGN1990 contain an additional class of greenhouses. These classes were reclassified into six land cover categories, specifically, pasture, cropland, built-up area, greenhouses, water and nature area. For the actual analyses, only the first three land cover types are used. The reclassification is shown in Table 1.

4 Methods

4.1 Extracting the Cities

In the land-cover maps, the built-up area category contains the entire infrastructure; therefore, various features such as settlements and roads are all represented by this class. As we are interested in evaluating the land cover configuration around urban areas, we extracted the cities from the built-up area pattern using the following GIS operations:

- Calculate built-up density: For each cell **c** in a land-cover raster, we calculated the fraction **Fc** of built-up area cells in its circular shaped neighbourhood of radius **r** (implemented using the Focal Statistics tool of ESRI ArcGIS 10).
- Create raster of high density clusters: we calculated a raster **S** wherein each cell for which **Fc** > **p** was assigned value 1 and **null** otherwise (implemented using the Raster Calculator tool).
- Erase holes in clusters: we updated **S** by assigning a value of 1 to all cells within the clusters (implemented using the Raster To Polygon, Feature To Polygon and Dissolve tools).
- Join nearby clusters: clusters that were within **d** meters from one another were assigned the same ID (i.e. they were considered as one settlement). Distance

d between pair of clusters was measured as the smallest distance between their boundaries (implemented using the Buffer tool).

We found by visual assessment that, with $\mathbf{p} = 0.4$, $\mathbf{r} = 250$ m (i.e. 5 cells) and $\mathbf{d} = 500$ m, we obtained clusters that well represented the settlements. We considered every cluster of cells with a unique ID as a single settlement.

4.2 Land-Cover Distributions Around Cities

We have selected the 20 largest cities based on their area in 1900 and present the individual distributions of the largest five. As these five cities are all situated on Holocenic deposits (characterized by soils that are prone to subsidence), we selected five additional cities from the largest 20 that are located on the Pleistocenic deposits (providing a more solid basis). These additional five were selected because land subsidence has often been an important driver of land use change (i.e. conversion of cropland to grassland), which could bias our results.

For each individual city, we calculated the distribution of pastures, cropland and nature as a function of distance from the city. The water, built-up, and greenhouse land cover categories were omitted. We created zones of 150 m width around cities, up to a distance of 10,000 m, and computed the fractions of the three land cover categories within these zones. We repeated this for the 1960 and 1990 maps. In the results section, we display the distribution of land cover as a function of distance for the individual cities and for all of the cities combined.

4.3 Static Land Cover Distributions

The observed changes in the land cover around the cities may have resulted from the outcome of two processes: (a) changes related to the expansion of the cities. For example, as cities expand, they "consume" the surrounding areas; this could change the land cover distribution around the city even if the land cover remained unchanged; and (b) actual changes in the land cover around the cities.

To estimate the contribution of the first process, i.e. that the land cover distributions are due to urban expansion while the landscape remains relatively stable, we constructed a new set of land cover distributions which depicted how the distribution would be if the cities expanded while the land cover remained static. We accomplish this by superimposing the city area at a given time onto the (surrounding) land cover map of a previous time period. Thus, for the individual cities, we construct two graphs of land cover distribution by distance based on: (a) the city delineations of 1960 and the land cover of 1900 and (b) the city delineations of 1990 and the land cover of 1960. We refer to these as the "static land cover distributions".

5 Results

5.1 Global Trends

The overall distribution of land cover areas for the different years is presented in Fig. 1. The built-up area increased from approximately 0.5 to 3.8 million hectares during the period of 1900–1990. This expansion occurred primarily at the expense of agricultural land, as this land was in the immediate surroundings of the cities. Subsequently, agricultural land expanded at the expense of natural areas. In the first period between 1900 and 1960, the natural area was reduced by more than 50 %, mainly by conversions to agriculture. This trend ceased in the second period, during which agricultural land again decreased, now not only by conversions to built-up land but also by conversions to new natural areas. Note that the total area increased due to the reclamation of land from the IJssel Lake: the Flevo Polders.

5.2 The Distribution of Land-Cover Around Cities

The land cover maps of Amsterdam, Utrecht and Leeuwarden (arbitrary examples) are presented in Fig. 2 for the 3 years. The maps of Amsterdam and Utrecht demonstrate rapid urban growth: the urban expansion between 1900 and 1960 appears relatively compact while a pattern of scattered development seems to emerge by 1990. In contrast, the growth of Leeuwarden remained compact.

The maps of Utrecht and Leeuwarden suggest that patches of cropland had been located near the cities' boundaries in 1900 and 1960, while this tendency is not visible in the 1990 map. In regard to Amsterdam, it appears as if the agricultural and natural landscape remained stable during the urban expansion.

The land-cover distributions around the five largest cities are shown in Fig. 3. Similar to the global fractions (Fig. 1), pasture was the dominant land cover for nearly the entire distance interval while there is no clear dominance of either nature or cropland. In The Hague, for example, nature was more abundant than cropland nearer to the city while cropland became more dominant at larger distances. In 1900 (Fig. 3a), we can observe a rapid change in the land cover composition near the cities. The fraction of pasture increased rapidly with distance from cities (while cropland decreased) until a distance of approximately 2 km. This tendency is discernible for four of the five cities; Rotterdam is an exception where the opposite is evident. The fraction of nature did not exhibit consistent qualitative trends with distance from the five cities.

In 1960 and 1990, the gradients in land cover fractions near the cities became weaker (Fig. 3b, c). Around Utrecht, for example, in 1900, the fraction of cropland decreased by 19 % within the first 2 km; in 1960, the decrease was approximately 15 %; while, in 1990, it was only 3 %. A similar tendency existed around The Hague and Leeuwarden. Around Amsterdam, the decrease in the percentage of

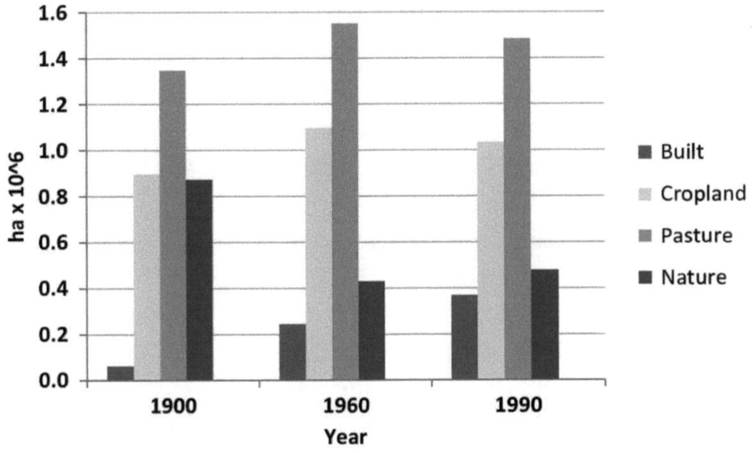

Fig. 1 The distribution of land cover areas in the Netherlands for 1900, 1960 and 1990

cropland had entirely ceased by 1960. This cessation is related to a large-scale conversion of cropland to pasture land mainly due to the subsidence of the peat soils.

The five largest cities were located on Holocenic deposits. The land-cover distribution around the other five cities, i.e. those situated on Pleistocenic deposits, is presented in Fig. 4. In 1900, there was more cropland around the cities on Pleistocenic deposits compared to the cities located on Holocenic deposits. By 1960, pasture also became more prevalent in the Pleistocenic regions.

Similar to the "Holocenic cities", in 1900, we can detect a gradient in the land cover composition near the "Pleistocenic cities" where the fraction of cropland is decreasing, and the fraction of pasture is increasing with distance. In 1960, this tendency was only discernible for Nijmegen and Arnhem and disappeared completely by 1990. The surroundings of Maastricht did not exhibit clear gradients for the entire period.

The fractions of the three land cover categories as a function of distance were averaged for the 20 largest cities (Fig. 5). The tendencies observed for the individual cities are also discernible in the averages. For 1900, the fraction of cropland decreased from approximately 38 % at a distance of 75 m to about 17 % at a distance of 2 km. This tendency was weaker in 1960: the fraction of cropland decreases from approximately 29 % at a distance of 75 m to about 18 % at a distance of 2 km. For 1990, this behaviour can no longer be observed: the fraction of cropland was approximately 20 % at a distance of 75 m and about 23 % at a distance of 2 km. We should note that the variation in land-cover fractions between cities is large. For example, the standard deviation for cropland is about 25 %, thus cities are quite different from each other with respect to the land cover distributions. However, despite the pronounced variation, the behaviour we described is still evident.

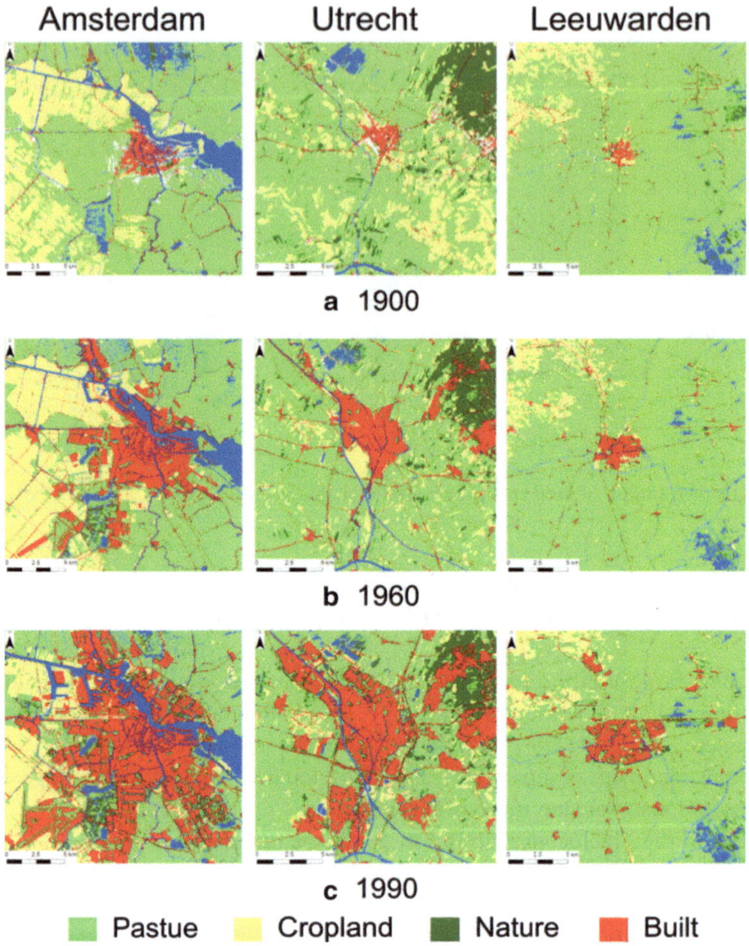

Fig. 2 The land cover configuration around Amsterdam, Utrecht, and Leeuwarden in 1900, 1960, and 1990

5.3 Static Land Cover Distributions

The static land cover distributions for the five largest cities are depicted in Fig. 6. The thick curve in Fig. 6a represents the land cover distribution in 1900 as measured from the city boundary of 1960; while the thin curve represents the land cover distribution of 1960 as measured from the city boundary of 1960 (as already shown in Fig. 3b). In Fig. 6b, the thick curve represents the land cover distribution in 1960, measured from the city boundary of 1990; while the thin curve represents the land cover distribution of 1990, measured from the city boundary of 1990 (as previously shown in Fig. 3c). Hence, deviations between the thick and the thin lines are due to land cover change around the cities without the

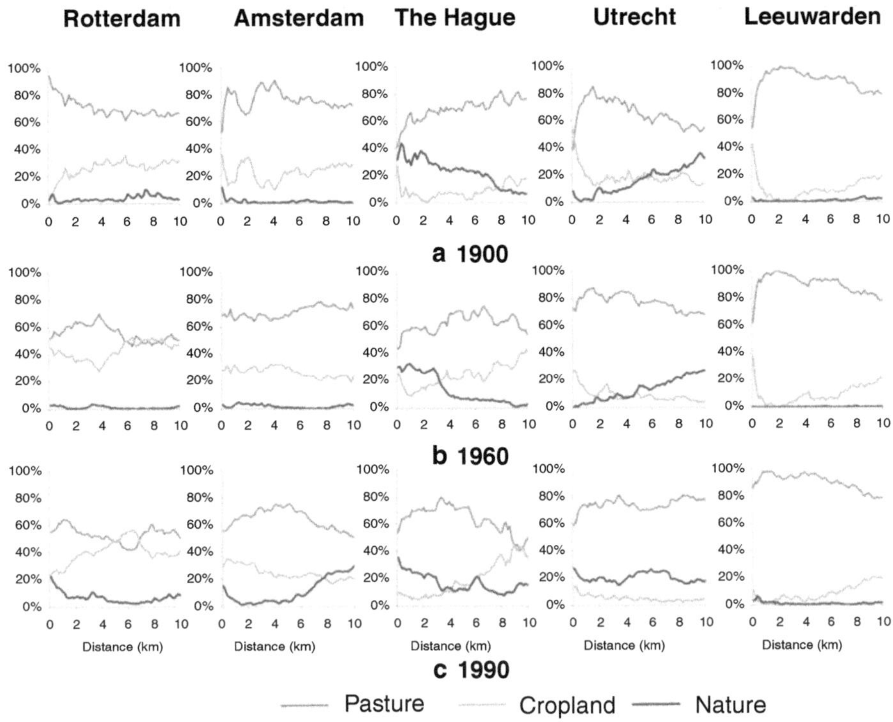

Fig. 3 The fraction of pasture, cropland and nature by distance from the urban area of the five largest cities (by area) in 1900

effect of the actual extension of the urban boundary itself. We can discern that the two lines are more similar between 1990 and 1960 than the lines of 1960 and 1900 which is most likely related to the different time intervals (60 years compared to 30 years). For the city of Leeuwarden, the resemblance between the two lines is most apparent: the curves for the two time intervals are almost identical apart from visible differences near the city boundary. Similar behaviours can be seen for Amsterdam (for both years) and for Utrecht and Rotterdam in 1990 (Fig. 6b).

Thus, it appears that the land cover distributions are affected by the cities at distances less than 2 km, but we do not observe any noticeable effect at larger distances. This is confirmed by the average static land cover distributions for all 20 cities (Fig. 7). The graphs for the cities of 1990 (Fig. 7b) indicate no clear gradient near the cities' edges. This is similar to the behaviour of the actual average distributions of 1990 (Fig. 5c). However, for the cities in 1960, we do discover a difference: the average static land cover distribution has only a minimal gradient near the cities' edges (Fig. 7a) while the empirical distribution (Fig. 5b) has a clear gradient. This confirms that the transformations we observe near the cities in 1960 are, indeed, related to land-cover change while, at larger distances, the land-cover was quite stable. In order words: It appears that cities in the Netherlands used to

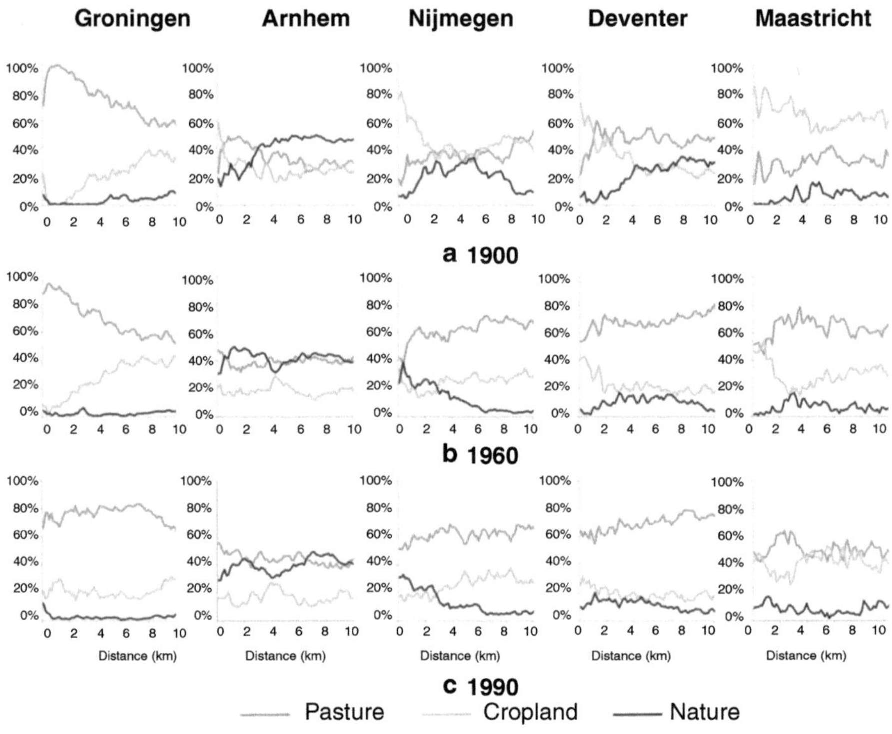

Fig. 4 The fraction of pasture, cropland and nature by distance from the urban area of cities located on Pleistocenic deposits

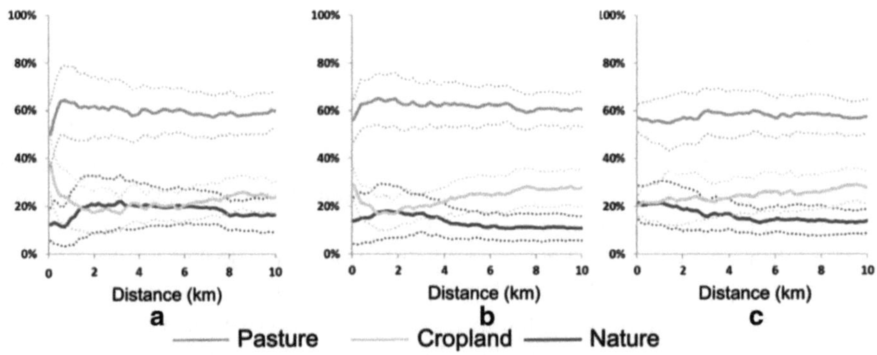

Fig. 5 The average fraction of pasture, cropland and nature by distance from the 20 largest cities (as in 1900): (**a**) 1900, (**b**) 1960, (**c**) 1990. The *dashed lines* indicate a one standard deviation interval

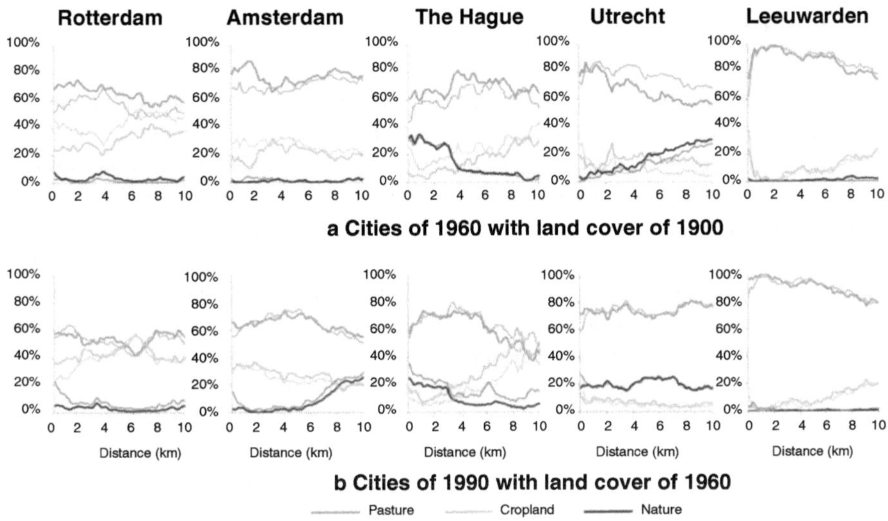

Fig. 6 The fraction of pasture, cropland and nature around cities: (**a**) the distribution based on 1960 cities and land cover of 1900 (**b**) the distribution based on 1990 cities and land cover of 1960. The *thin curves* represent the actual fraction as depicted in Fig. 3

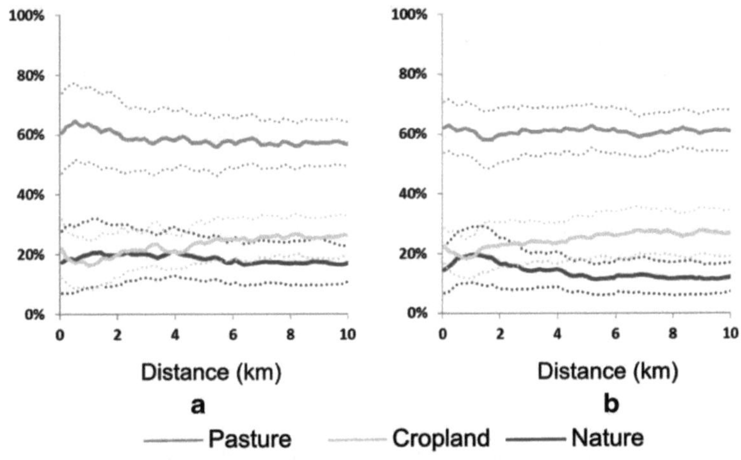

Fig. 7 The average fraction of pasture, cropland and nature for (**a**) 1960 cities and land cover of 1900 (**b**) 1990 cities and land cover of 1960. The *dashed lines* indicate a one standard deviation interval

have a limited influence on the land cover distribution in their hinterland, this influence has diminished even further with time.

6 Conclusions

We found that the distribution of cropland, pasture and nature as a function of distance from cities varies considerably between the cities. However, for many cities, we did observe the following general behaviour: In the past, we could discern a clear trend of higher fractions of cropland near the cities (and lower fractions of pasture land). This tendency weakened by the middle of the century and almost completely ceased by 1990. This behaviour was also evident in the average distributions of the 20 largest cities despite the high variability. The higher fractions of cropland near the cities in 1900 and 1960 likely represent horticulture activities for vegetable production which were performed near and within the cities.

By superimposing cities onto older land cover patterns, we discovered that the landscape around the cities is relatively robust. Nevertheless, the analysis suggests that the land cover trends near cities cannot be explained by urban expansion within a static landscape. That is, the composition of land cover near cities in 1900 and 1960 were likely influenced by the proximity to the cities. For 1990, a clear influence of cities could no longer be observed.

The analysis presented in this chapter was limited by the quality of the dataset. The dataset has good spatial resolution, but the number of land cover categories is quite small. For this reason, we could not explore non-agricultural activities near the cities nor provide a more detailed account of the agricultural activities (i.e. we could not distinguish between gardening and arable cultivation or between intensive dairying and rough grazing). Despite these limitations, spatial and temporal trends remained visible. With the increasing availability of historical datasets in the coming years, we can expect more comprehensive and detailed studies of the spatial impact of past urbanization.

Acknowledgements Erez Hatna acknowledges support from Joshua M. Epstein's NIH Director's Pioneer Award, number DP1OD003874 from the National Institutes of Health.

References

Alltera, (2012). *Historisch Grondbruik Nederland.* http://www.alterra.wur.nl/NL/Producten/GIS-bestanden/HGN/. Accessed 06 Sept 2012.

Alo, C. A., & Pontius, R. G. (2008). Identifying systematic land cover transitions using remote sensing and GIS: The fate of forests inside and outside protected areas of Southwestern Ghana. *Environment and Planning B, 35*(2), 280–295.

Angel, S., Parent, J., Civco, D. L., & Blei, A. M. (2012). *Atlas of urban expansion.* Cambridge, MA: Lincoln Institute of land policy.

Bakker, M. M., Hatna, E., Mucher, S., & Kuhlman, T. (2011). Changing environmental characteristics of European cropland. *Agricultural Systems, 104*, 522–532.

CBS. (2012). Population, households and population dynamics; from 1899. Centraal bureau voor de Statistiek. Online document available at statline.cbs.nl/statweb/

Grigg, D. B. (1984). *An introduction to agricultural geography.* London: Hutchinson Educational.

Knol, W. C., Kramer, H., & Gijsbertse, H. (2004). *Historisch Grondgebruik Nederland: Een landelijke reconstructie van het grondgebruik rond 1900*. Alterra: Wageningen.

Rhind, D., & Hudson, R. (1980). *Land use*. London: Methuen.

Stone, K. H. (1975). Geographical effects of urbanizing on ruralizing in Europe, 1920–2000. *European Demographic Information Bulletin, 6*(1), 2–22.

Van der Valk, A. J., & De Vries, J. (1996). De Rijksplanologische Commissie: in de schemerzone tussen planning en politiek. *Stedebouw & Ruimtelijke Ordening, 77*(5), 24–31.

Verburg, P. H., & Overmars, K. P. (2009). Combining top-down and bottom-up dynamics in land use modeling: Exploring the future of abandoned farmlands in Europe with the Dyna-CLUE model. *Landscape Ecology*, *24*, 1167–1181.

von Thünen, J. H. (1826). Der Isolierte Staat in Beziehung auf Landschaft und Nationalökonomie. English edition: Thünen, von J. H. (1966). Von Thünen's Isolated State (trans: Wartenberg C.M.). Oxford: Pergamon Press.

White, R., & Engelen, G. (2000). High-resolution integrated modelling of the spatial dynamics of urban and regional systems. *Computers, Environment and Urban Systems, 24*(5), 383–400.

Investigating Land-Use Dynamics at the Periphery of a Fast-Growing City with Cellular Automata at Two Spatial Scales

Danielle J. Marceau, Fang Wang, and Nishad Wijesekara

Abstract The city of Calgary in southern Alberta, Canada has a current population of 1,120,000 inhabitants; it has experienced steady population and unprecedented land-cover growth over the past six decades due to the strong Alberta economy centered on petroleum industry, tourism, and agriculture. The residential land developments currently expanding at the periphery add up to a sprawling city that is expected to reach 1.5 million inhabitants in 2020 and 2.3 million over the next 50–70 years. Such a rapid growth imposes dramatic land-use changes within and around the city that affect the adjacent environmentally sensitive areas, particularly in the western fringe of Calgary, located in the Elbow River watershed. This paper describes the application of two cellular automata (CA) designed at two spatial scales to investigate the land-use dynamics occurring respectively in the whole watershed and in the eastern portion of the watershed, immediately adjacent to the City of Calgary. The first CA is a cell-based model applied at a spatial resolution of 60 m while the second one is a patch-based model specifically designed to account for the greater level of details that can be observed at the resolution of 5 m. These two models were used to simulate different land development scenarios over a period of 20–30 years. The CA model implemented at the scale of the watershed provides useful information about alternative spatial distributions of urban areas that can occur according to spatial constraints imposed on land development, which can help decision makers find the most adequate distribution of population in regards to the resources that are available to sustain that population in the watershed. The CA model designed at fine spatial resolution allows the identification of detailed land-use classes and changes in their internal structure. It reveals how land consumption can be considerably diminished by encouraging the protection of sensitive areas and increasing the density of existing and new urban residential areas.

D.J. Marceau (✉) • F. Wang • N. Wijesekara
Department of Geomatics Engineering, University of Calgary, 2500 University Dr. NW, T2N 1N4 Calgary, AB, Canada
e-mail: dmarceau@ucalgary.ca; wanf@ucalgary.ca; gnwijese@ucalgary.ca

D. Malkinson et al. (eds.), *Modeling of Land-Use and Ecological Dynamics*, Cities and Nature, DOI 10.1007/978-3-642-40199-2_4, © Springer-Verlag Berlin Heidelberg 2013

1 Introduction

Land-use activities that involve the conversion of natural landscapes for human use or the intensification of management practices have transformed a large proportion of the planet's land surface with major consequences on climate, hydrology, forest resources, biotic diversity, ecosystem services, air quality, and land degradation (Foley et al. 2005; Rindfuss et al. 2004). Human transformations of ecosystems and landscapes affect the ability of the biosphere to sustain life at a time when humanity has become increasingly dependent on its resources (Foley et al. 2005). The need for understanding land-use changes in order to improve land management, assess future trends, and ensure environment sustainability has given rise to the novel discipline of land-use science (Gutman et al. 2004) along with international scientific initiatives such as the Land-Use and Land-Cover Change and the Global Land Projects of the International Geosphere-Biosphere Program and the International Human Dimensions Program (Verburg 2006; Veldkamp 2009).

A major cause of landscape transformation is urbanization, i.e., the physical growth of urban areas due to rural migration and population concentration into cities. Like most urban centers in North America, the city of Calgary in southern Alberta, Canada, has experienced steady population and unprecedented land-cover growth over the past six decades due to the strong Alberta economy, natural increase, and net migration (City of Calgary 2009). Since 1988, the population has increased from 657,118 to 1,120,225 inhabitants in 2012, an increase of 70 % (Civic Census Results City of Calgary 2012). Calgary is currently the fourth-largest city in Canada and one of the most attractive labor markets in the country, with economic activities centered on petroleum industry, tourism, and agriculture. It is predicted that its population will reach 1.6 million by 2037 and just fewer than two million inhabitants by the twenty-second century (City of Calgary 2009).

The rapid increase of population imposes dramatic land-use changes within and around the city that affect the adjacent environmentally sensitive areas. In particular, the western fringe of Calgary, located in the Elbow River watershed at only 80 km from Banff National Park, is under intensive pressure for land development. Since 1989, over 1,000 lots have been created for residential purposes, 65.8 % of them being 2–4 acres in size (Central Springbank Area Structure Plan 2001). Such low density and non-contiguous development also occur further west in the watershed, causing loss of productive agricultural lands, forest cover, surface water bodies, and increasing levels of water pollution. Of particular concern is the sustainability of water resources in the watershed that supplies about 40 % of Calgary's drinking water (Parks Foundation Calgary 2008) and fulfills water demands from industry, agriculture, golf courses, oil and gas, timber harvesting, recreation, and residential and commercial use. Existing and projected changes in population and land use, combined to climate might seriously compromise water availability and quality in the near future (Wijesekara et al. 2012; BRBC 2010; Chen et al. 2006; ERWP 2012). Therefore, there is a growing need from scientists and planners to understand current and future land use in order to balance land development with sustainable growth, and prevent unintended environmental problems.

This research has been undertaken in collaboration with the Calgary Regional Partnership, the Rocky View County, and Alberta Environment Sustainable Resource Development to understand the historical land-use changes in the Elbow River watershed and explore scenarios of land development over the next 20–30 years using two cellular automata (CA) models designed at two spatial scales: the whole watershed and a small eastern sub-area immediately adjacent to the city of Calgary. The first CA is a cell-based model applied at a spatial resolution of 60 m, while the second one is a patch-based model specifically designed to account for the greater level of details that can be observed at the resolution of 5 m. The fine-scale CA can be seen as a tool for *zooming in* on a small area of the watershed where intense land development occurs. The two models are then used to investigate the main land-use transitions and driving factors of change and interpret the land-use patterns that evolve through time according to different scenarios of development.

In the remaining sections, the two study areas are presented, followed by a detailed description of the CA models that were developed and the scenarios that were simulated. The interpretation of the results emphasizes the different perspectives about the land-use dynamics revealed by the two CA models in terms of the main land uses and transitions, the factors driving the changes, the land-use intensification, the fragmentation of the landscape, and the impact of land development scenarios on agriculture, forest, and patches of natural land.

2 Methodology

2.1 Description of the Two Study Areas

The Elbow River watershed in southern Alberta drains an area of 1,238 km^2 (Fig. 1). From west to east, 65 % of the watershed is located in the Kananaskis Improvement District in the Canadian Rockies, while the remaining area is divided among the rural Rocky View County (20 %), the Tsuu T'ina Nation (10 %), and the City of Calgary (5 %). The watershed is the source of the Glenmore reservoir, which fulfills part of the drinking water supply to Calgary. In addition to urban areas (6 %) that are concentrated in the north-east portion of the watershed, agriculture and rangeland/parkland represent respectively 17 % and 6 % while evergreen and deciduous forests respectively cover 34 % and 10 %. Recent clear-cuts can be observed in about 2 % of the western portion of the watershed. The Tsuu T'ina Nation in the south-east and the large rocky area in the west remain undeveloped.

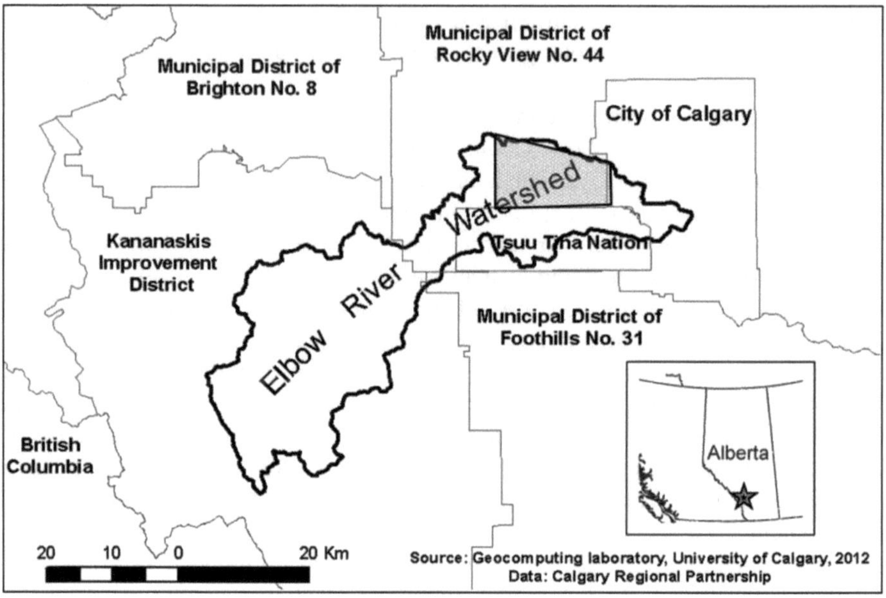

Fig. 1 Elbow River watershed and second study area (shown by the polygon on the *top right*)

In addition to the urbanization occurring in the area adjacent to the City of Calgary (described below), intensification of land development also occurs in Bragg Creek, a hamlet located 30 km west of Calgary at the confluence of the Elbow River and Bragg Creek, that offers mountain scenery and numerous recreational activities. The watershed also includes the unique community of Redwood Meadows, a townsite that is located at about 10 km of Bragg Creek within the Tsuu T'ina Nation Indian reserve. Along with Calgary, these two small communities create a corridor of development, facilitated by the presence of highways that connect them.

The second study area is a small subset in the eastern part of the watershed that covers 250 km^2 (Fig. 1). About 14 % is located within the boundary of the city of Calgary while the remaining 86 % is located in the Rocky View County. This area contains a unique and diverse mix of natural features and rural and urban components. The main land covers/land uses consist in agriculture (63 %, including rangeland and cropland), forest (12 %), and built-up areas (25 %). This last class includes low density residential (density is usually lower than 0.5 unit per acre), high density residential (density is usually greater than 6 units per acre), commercial, and recreational areas (i.e., golf courses). Moreover, three major highways (the Trans-Canada Highway 1 at the north, Highway 22 at the west, and Highway 8 at the south) and the Elbow River cross the area and influence its land use. Historically, this area was considered a productive and socially vibrant agricultural area. However, with its convenient proximity to Calgary, it has undergone dramatic land-use changes over the last decade and is under considerable pressure for further development, particularly along the three highways (Rocky View County/City of Calgary Intermunicipal Development Plan 2011).

2.2 *The Cell-Based CA Model Designed for the Whole Watershed*

A detailed description of the cell-based CA designed at the scale of the whole watershed can be found in Hasbani et al. (2011). The calibration and validation of the model were done using historical land-use maps produced from Landsat Thematic Mapper imagery acquired during the summers of 1985, 1992, 1996, 2001, 2006 and 2010 at the spatial resolution of 30 m. These maps contain nine dominant classes: water, rock, roads, agriculture, deciduous, evergreen, rangeland/parkland, urban areas, and forest clear-cuts. The quality of these maps was assessed through field verification and the use of third-party maps (from Google). They were further revised to ensure their temporal consistencies using a computer program written in IDL® (Wijesekara et al. 2012). The land-use map of the year 2010 is shown on Fig. 2.

An exhaustive sensitivity analysis was conducted to identify the best configuration of the model. It revealed that a cell size of 60 m, a neighborhood consisting of three concentric rings of 5, 9, and 17 cells (corresponding to 300, 540, and 1,020 m respectively), and four external driving factors (distance to the Calgary City center, distance to a main road, distance to a main river, and the ground slope) were the most appropriate to capture the land-use patterns in the watershed. The particular neighborhood configuration composed of the three concentric rings was selected to take into consideration the local and extended influence on the central cell and to reduce the bias from distant cells. Four maps representing each external driving factor were built using the Euclidian distance function available in ArcGIS.

Eight land-use transitions were considered for the simulations (Table 1).

The transition rules were extracted using the historical land-use maps and the maps corresponding to the four external driving factors. The extraction procedure involves the following steps. First, for each land-use transition, the cells that have changed state in the historical maps are identified along with the number of cells of a particular state in their neighborhood and the values of each external driving factor. This provides information about the conditions that prevailed around each cell that has changed state. This information is displayed through a graphical user interface (GUI) in the form of frequency histograms that can be interpreted by the modeler. An example of a frequency histogram displaying the occurrence of cells that have changed from agriculture to built-up between 1985 and 2006 considering their distance to roads is provided in Fig. 3. This histogram reveals that the largest number of cells that have changed state from agriculture to built-up during that period were located between 100 and 600 m from a main road. This range of values

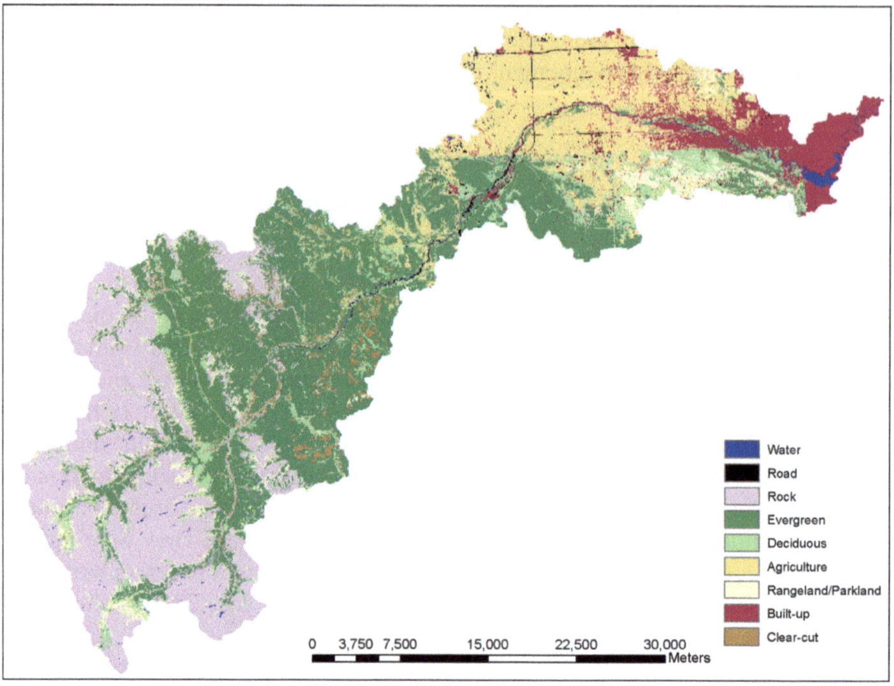

Fig. 2 Land-use map of 2010

Table 1 Land-use transitions considered during the simulations

From	To
Evergreen	Agriculture
	Built-up
Deciduous	Agriculture
	Built-up
Agriculture	Rangeland/parkland
	Built-up
Rangeland/Parkland	Agriculture
	Built-up

can be selected by the modeler to become the parameter values of the conditional transition rules of the model that will take the form of:

If distance to a main road is between 100 and 600 m and number of built-up cells within the first neighborhood ring is between 0 and 17 and distance to downtown Calgary is between 1,000 and 2,200 m then the central cell might change from Agriculture to Built-up area.

All possible transition rules are created by combining the identified ranges of values from the histograms (Hasbani et al. 2011).

The second step involved in the extraction of the transition rules is to translate the conditional rules into mathematical rules. This allows the use of a quantitative

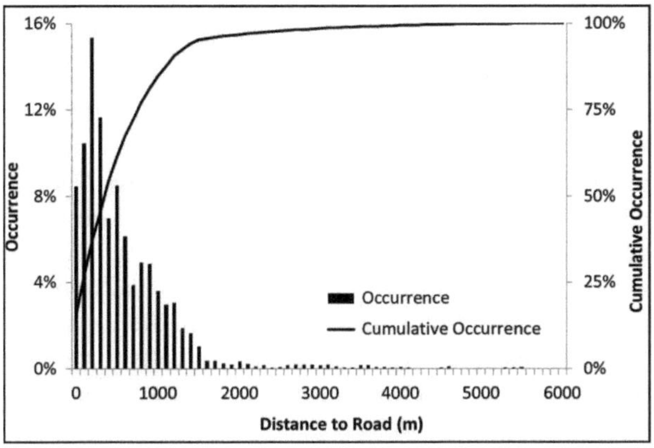

Fig. 3 Histograms illustrating the occurrence of cells that have changed state from agriculture to built-up between 1985 and 2006 considering their distance to main roads

index, referred to as the resemblance index (RI) to determine if a cell will change state or not. The mathematical rules are built by calculating the mean and standard deviation of the defined ranges of values on the frequency histograms, which will become the coefficients of the parameters of the mathematical rules. Using the coefficients of each transition rule, the resemblance index is calculated using Eq. 1, which quantitatively describes the similarity between the conditions prevailing in the neighborhood of a cell at the time step of the simulation and the conditions observed in the neighborhood that have been used to generate the values of the parameters of the transition rule.

$$RI = \sum_{i=1}^{m} \frac{|n_i - \overline{x}_i|}{\sigma_i} \tag{1}$$

where m is the number of layers (corresponding to the number of external driving factors plus the number of land-use classes multiplied by the number of neighborhood rings), n_i is the value in layer $i, \overline{x}_i$ is the mean value for layer i in the transition rule, and σ_i is the standard deviation for layer i in the transition rule. If the standard deviation is zero for layer i, then $\frac{|n_i - \overline{x}_i|}{\sigma_i} = 0$ if $n_i = \overline{x}_i$ or otherwise equals positive infinity. Accordingly, $RI \in \Re^+$ and the smaller RI is, the more similar are the conditions surrounding a cell to the ones used to define the transition rule (Hasbani et al. 2011).

During the simulation, at each time step, the neighborhood composition of every cell is read and the level of correspondence with the parameters of the transition rules is computed. The cells having the highest level of correspondence based on user-specified constraints and the influence of each rule are subjected to change state. Decision on which cell should be associated to each type of land-use change is made by recursively sorting the type of land-use changes and selecting the cell having the smallest RI value. Once the required number of cells associated to each type of land-use change is met or when no more cells can be assigned, the model generates the new land-use map and updates the statistics that correspond to the percentage of cells associated to each rule and each type of change. If the numbers of cell associated to each rule and each type of land-use change is different than the numbers found from the historical data and previous time steps, a correction is applied at the next time step (Hasbani et al. 2011).

In all simulations, two local constraints were applied to respectively forbid new built-up development within the Tsuu T'ina Nation reserve and to restrict any changes in the forest reserves within the Kananaskis Improvement District. Simulations were conducted from 1985 to 2010. The quality of the calibration was evaluated by comparing the simulated land-use maps of 2006 and 2010 with the reference land-use maps generated from the Landsat TM images. The comparison was done using a neighborhood of five cells to capture the land-use patterns rather than the state of each cell at their exact spatial location (Hasbani 2008). A correspondence of 96 % and 91 % between the simulated maps and the reference maps was obtained for the years 2006 and 2010, respectively. This high correspondence reflects the fact that change was limited in some parts of the watershed.

2.2.1 Scenarios to Forecast Land-Use Changes in the Watershed

Four scenarios were simulated with the cell-based CA (Table 2): the business-as-usual scenario (BAU) that assumes that conditions observed in the past will still prevail in the future, a scenario with a new centralized development plan within the Rocky View County (RV-LUC), a scenario with a new centralized development in the area of Bragg Creek (BC-LUC), and a scenario where land-use is changed based on the forecasted population trends according to Calgary Economic Development (2010) (P-LUC). For all scenarios, the simulations were carried out for the years 2016, 2021, 2026, and 2031 using the initial land-use map of 2010. Using forest harvesting data obtained from Alberta Environment, clear-cut areas relevant to each future year (2016, 2021, 2026, and 2031) were overlaid on each simulated land-use map in order to take into account possible changes in the forested portions of the watershed. The results of the simulation are presented in Sect. 3.1.

2.3 The Patch-Based CA Model Designed for the Watershed Sub-Area

To tackle the challenge of simulating land-use changes at the fine resolution of 5 m, a novel patch-based CA model was designed in which the real-world entities (i.e., the detailed land-cover/land-use classes) are represented as patches. A patch refers to a collection of adjacent cells that, when combined together, represent an entity differing from its surroundings in nature or appearance (Weins 1976; Fu and Chen 2000). In this patch-based CA model, the transition probabilities of the cells in their respective states and locations within each patch are calculated to account for the

Table 2 Simulated land-use change scenarios

Scenario	Description	Applied constraint
BAU	Business as usual, following trends detected from the historical land-use maps	N/A
RV-LUC	New centralized development within the Rocky view county	Local spatial constraint to promote new urban development
BC-LUC	New centralized development in the area of Bragg Creek	Local spatial constraint to promote new urban development
P-LUC	Development based on projected population growth	Global constraint set based on population growth for each future year

internal heterogeneity of the entities being represented. A detailed description of the model can be found in Wang and Marceau (2012).

To calibrate the patch-based model, four historical land-use maps were generated from SPOT-5 multi-spectral (MS) and panchromatic (PAN) remote sensing images acquired at the respective spatial resolution of 10 m and 5 m for the years of 2003, 2006, 2008, and 2011. An image fusion procedure was applied using ENVI 4.6 (Exelis Visual Information Solutions 2012) to combine the MS and PAN images at 5 m resolution. An object-based classification procedure was performed using the software *eCognition* (Trimble 2012) to identify 14 land-use/cover classes, which are listed in Table 3. The classification accuracy was assessed through field validation using 255 points distributed over the study area. Temporal inconsistencies among the historical maps were further identified and removed. The land-use map of the year 2011 is shown as an example (Fig. 4). The 14 land-use classes were sub-divided into two categories: dynamic classes and static classes. Dynamic classes include country residential, urban residential, forest, agriculture, and under development; these classes will change during the simulation. The remaining of the classes are static, which means that they do not change over time, but have a high influence on the dynamic classes.

Two sets of factors driving land-use changes in the study area were considered: internal factors (i.e., neighborhood influence defined by the cell states, extracted from the historical land-use maps) and external factors (i.e., extracted from external data sources). To determine the internal factors, a neighborhood composed of three concentric rings (Hasbani et al. 2011) was used. Through an analysis of the historical land-use maps, the *average radius* (AR) within patches was calculated for each land-use class along with the average distance between different land-use patches. Then, the three neighborhood ring radii were defined as: (i) the minimum AR (115 m), (ii) the maximum AR (385 m), and (iii) the average distance between patches (825 m). The external factors that were considered include slope, distance to road level 1 (Highway 1), distance to road level 2 (Highway 8 and Highway 22), distance to road level 3 (other main roads in the study area), and distance to Calgary's boundary. Such accessibility factors have been proven to be important in many previous land-use CA simulations (Dietzel and Clarke 2007; Liu et al. 2010).

Table 3 Land-use classes identified at 5 m resolution

	Number	
Static land-use class	1	Commercial/Institutional
	2	Industrial
	3	Airport
	6	Road level 1
	7	Road level 2
	8	Road level 3
	9	Golf course
	11	Water
	13	Green area
Dynamic land-use class	4	Urban residential (UR)
	5	Country residential (CR)
	10	Agriculture (A)
	12	Forest (F)
	14	Under development (U)

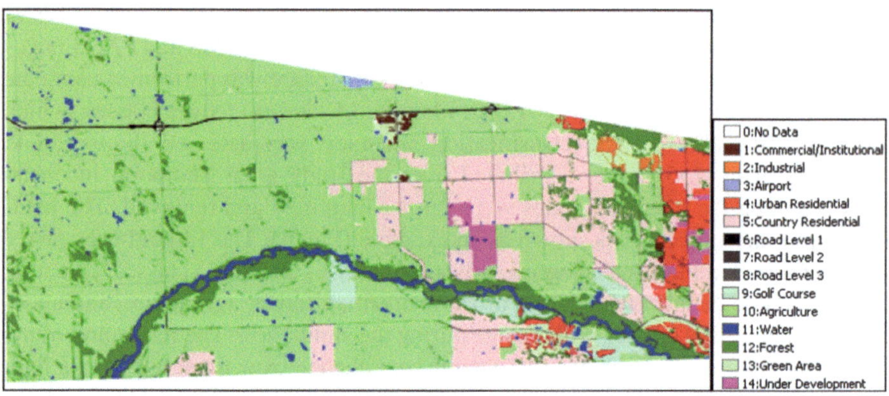

Fig. 4 Historical land-use map of the year 2011

Rather than conducting a sensitivity analysis to identify the best combination of internal and external driving factors, a data mining technique called Rough Set Theory (RST) (Pawlak 1982) was applied to facilitate the factor reduction and selection for the calibration of the model. This technique derives an optimal set of factors from an original set while minimizing the redundancy and retaining the original factors. Following the methodology developed by Wang et al. (2011), RST was applied to select the driving factors that are listed in Table 4.

Five main land-use transitions were considered based on observations made from the historical land-use maps: (i) from country residential (CR) to under development (U), (ii) from agriculture (A) to under development (U), (iii) from forest (F) to under development (U), (iv) from under development (U) to country residential (CR), and (v) from under development (U) to urban residential (UR).

Table 4 Selected factors using rough set theory

Initial land-use class	RST selected factors
CR	Dist2cityBdry; Dist2roadL1; Dist2roadL2; Dist2roadL3; NH0#4; NH0#5; NH0#10; NH0#12; NH1#4; NH1#5; NH1#8; NH1#10; NH1#11; NH2#4; NH2#7; NH2#8
A	Dist2cityBdry; Dist2roadL1; Dist2roadL2; Dist2roadL3; Elevation; NH1#4; NH1#5; NH1#8; NH1#10; NH1#12; NH2#11; NH2#12; NH2#14;
F	Dist2cityBdry; Dist2roadL1; Dist2roadL2; Dist2roadL3; Elevation; NH0#5; NH0#12; NH1#8; NH1#10; NH1#11; NH1#12; NH1#14;
U	Dist2cityBdry; Dist2roadL1; Dist2roadL2; Dist2roadL3; NH0#4; NH0#5; NH0#14; NH1#4; NH1#5; NH1#8; NH1#10; NH1#14; NH2#4; NH2#5; NH2#8; NH2#10; NH2#14;

The approach used to define the transition rules is the transition probability function, which has been widely used due to its simplicity, clarity of the parameters, and easy interpretation of the results. The general format of a transition probability function is defined as:

$$S_{xy}^{t+1} = f\left(P_{xy}^{t}\right) \tag{2}$$

where, S_{xy}^{t+1} is the land-use state in the location (x, y) at time step $t + 1$, and P_{xy}^{t} is the transition probability of state S in location (x, y) at time step t. According to transition probability rules, the change of cell's state is based on the transition probability value of the cell at that location. The higher the transition probability value is, the higher the chances are for the transition to occur. In this study, the Weight of Evidence (WOE) method was chosen to calculate the transition probability function. Originating from Bayesian theory, WOE can be used to combine information of a set of categorical factors from known locations and quantify the transition probability of the unknown locations. This method has been proven to be robust even with small sample sizes, it does not rely on the assumption that the input data are normally distributed, and the interpretation of the weight values is very intuitive (Dickson et al. 2006). Moreover, Almeida et al. (2003) demonstrated that WOE can accommodate the nonlinear characteristics of a complex urban system.

The application of WOE for calculating the transition probability follows the methodological procedure described by Bonham-Carter et al. (1989). The posterior probability of a transition (i.e., $s \rightarrow k$) is calculated using the function defined as:

$$P\{s \rightarrow k | B \cap C \cap D \ldots \cap N\} = \frac{e^{\sum W_{N^+}}}{1 + e^{\sum W_{N^+}}} \tag{3}$$

where B, C, D, and N are the values of the factors,

$W_N{}^+$ represents the weight for each factor.

The weights of the factors are assigned based on the prior probability of the known locations.

$$W^{+} = Ln\left(\frac{P(E_i|D)}{P(E_i|\overline{D})}\right) \tag{4}$$

$$P(E_i|D) = \frac{N(E_i|D)}{N(E_i|T)} \tag{5}$$

where $N(E_i|D)$ is the number of the occurred transitions under the presence of factor E_i,

$N(E_i|T)$ is the total number of the cells under the presence of factor E_i, and

$P(E_i|D)$ is the prior probability of the transition under the presence of factor E_i.

Since WOE only applies to categorical data, an important step is to break the values of each factor into a series of ranges (i.e., categories), following the method described by Agterberg and Bonham-Carter (1990). First, the relationship curve between the values of the factor and the number of transitions that occurred in the historical data was constructed. Then, breaking points of the curve were determined by creating a best-fitting curve using a series of straight-line segments employing a line-generalizing algorithm. A weight was estimated for each range of each factor using the following function:

$$W^{+} = \ln\left(\frac{y_{n=k} - y_{n=k-1}}{A_{n=k} - A_{n=k-1}}\right) \tag{6}$$

$$y_n = A_n * \exp(W^{+}) \tag{7}$$

where n is the number of ranges for each factor, k represents the breakpoints defined for the ranges, and A_n is the number of cells for each range.

At each time step, a transition probability map for each land-use transition was calculated using the WOE function with the initial land-use map and the values of the selected factors as the input. It was observed in the historical maps that three types of transitions to the class under development happen simultaneously, namely from country residential, forest, and agriculture. A unique probability map was created by combining the transition probabilities of these three types of land-use transitions.

The simulation procedure was performed as follow (Fig. 5). First, patch information about the three final land-use classes (i.e., U, CR, and UR) was extracted from the historical land-use maps including mean patch size (MPS), minimum patch size (MiPS), number of changed cells (NCC), and number of patches (NOP). This information was used to create potential patches. For the transition to under development, the seeds of the patches corresponding to the cells having the highest values in the combined transition probability map were selected. For the transitions from under development to country residential and to urban residential, the seeds of the patches were selected from the original transition probability maps generated by

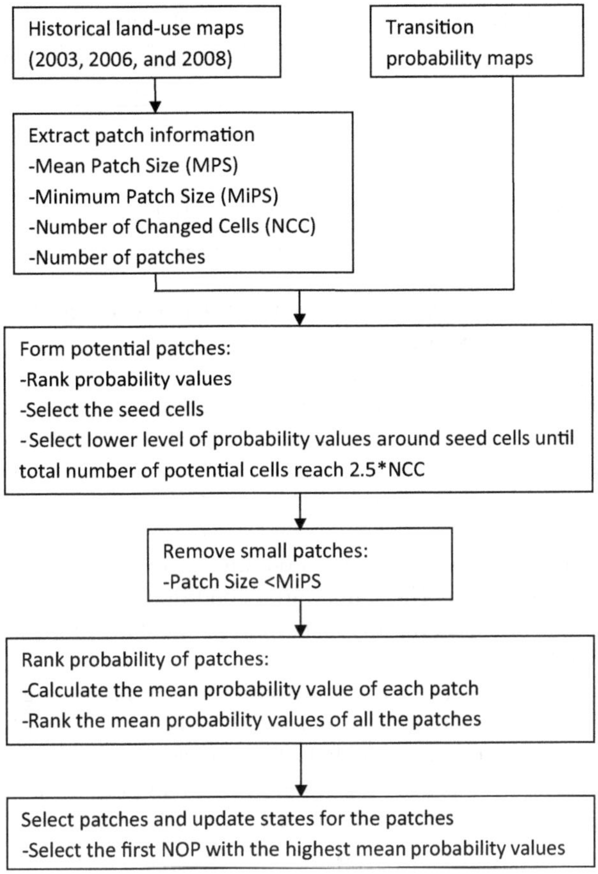

Fig. 5 Flow chart of the patch-based simulation procedure

the WOE method. Then the cells with a lower level of probability values around the seed cells were selected and merged to the seed cells to form patches. This process was repeated until the potential cells for the patch reaches a certain number that is close to 2.5 times of NCC found in the historical data. The number 2.5 was obtained during the calibration of the model. Third, small patches were removed based on the MiPS found in the historical maps. Fourth, a mean transition probability value was calculated for each patch afterwards. Finally, the mean transition probability values for the patches were ranked; the patches with higher mean probability values were selected until the number of patches equals to NOP. Finally the states of the selected patches were updated. This procedure was repeated for each type of land-use transitions resulting in a new land-use map created by combining all the changes.

Table 5 Parameters for the business-as-usual scenario

Land-use change	Number of changed cells	Minimum patch size	Number of changed patches
Change to under development	83,513	5,841	13
Change from under development to high density residential	41,711	5,845	7
Change from under development to low density residential	41,471	6,042	6

2.3.1 Scenarios to Forecast Land-Use Changes Using the Patch-Based CA Model

Based on information extracted from the local municipal development plan (Rocky View County 2012), three scenarios were designed: the *business-as-usual* scenario, the *protective growth* scenario, and the *smart growth* scenario. Simulations of these scenarios were run from 2011 to 2041 with a time step of 3 years.

The business-as-usual scenario assumes that the future land-use change rate remains unchanged when conditions are similar to the ones observed from the historical data. The same initial conditions as those used to simulate from 2003 to 2011 were used for this scenario. The input parameters for this scenario are listed in Table 5.

The protective growth scenario aims at promoting land development in a more sustainable way by considering watershed and agriculture protection while accommodating the same number of people as the one described in the business-as-usual scenario. To protect the water source and prevent development in areas located at proximity to the river, a buffer zone of 120–200 m wide was created from the center line of the river to cover not only the fresh water, but also the flood plain and the vegetation along the river, which are of great ecological value. Development that falls within this buffer zone was excluded.

Similarly, the sites having high agriculture capabilities were excluded from development. To implement this scenario, the land capacity for agriculture map from Canada Land Inventory (2008) was used. Land development was excluded on class 1 (no significant limitations) and class 2 (moderate limitations), considered as good quality agriculture land. The input parameters for the Protective Growth Scenario are listed in Table 6.

The smart growth scenario simulates *smart growth*, a contemporary urban planning that encourages the sustainable development of cities through a compact urban form in order to conserve more land (Plan It Calgary Workbook 2007). One way to conserve land, while allowing the same number of people to settle in the area is to increase land-use efficiency defined as the area of land consumed to accommodate a certain number of people. This scenario takes into consideration the projected population and economic growth in the city of Calgary and around, with the objective of increasing the land-use efficiency of the development by 30 % as suggested by Imagine Calgary Plan for Long Range Urban Sustainability (2006). To increase land-use efficiency, a major strategy is to increase high-density

Table 6 Parameters for the protective growth scenario

Change type	Number of changed cells	Minimum patch size	Number of changed patches
Change to under development	55,675	5,841	9
Change from under development to high density residential	44,116	5,845	7
Change from under development to low density residential	11,559	6,042	2

residential (i.e., Urban Residential) and decrease low-density residential (i.e., Country Residential), which then reduce the per capita demand for the occupied land.

To calculate the input parameters for this scenario, the population and dwelling statistics for the Rocky View County and the city of Calgary were used (Table 7). According to Statistics Canada (2012), the cumulative growth for the city of Calgary from 2006 to 2011 is 10.9 % while the cumulative growth rate for the Rocky View County is 9.9 %. Based on this information, an average population growth rate of 2 % per year was applied in the study area, which corresponds to about 2,643 people at each time step of the simulation (i.e., 3 years).

When dividing the population by the dwellings in 2011, for both the city of Calgary and the Rocky View County, each dwelling will have an average of three people. For the class Country Residential, each dwelling covers above 2 acres per unit, while for Urban Residential, it covers about 0.17 acres per unit. In other words, each person in Urban Residential occupies about 229 m^2 (corresponding to 9 cells in the CA model) and one person in Country Residential area occupies about 2,698 m^2 (corresponding to 108 cells). To accommodate a total growth of 2 % of population in the study area, different ratio of Country Residential and Urban Residential area can exist, which will result in different land-use efficiency.

Calculated from the historical data, the ratio of the current population settling in Urban Residential and in Country Residential is about 12:1. This ratio needs to be improved to 37:1 in order to increase the land-use efficiency by 30 %. To accommodate the projected population increase at each time step of the simulation, the number of changed cells and the number of changed patches for each type of land-use transition is listed in Table 8.

3 Results and Interpretation

3.1 The Whole Watershed

Since different land-use changes dominate in the east and west sub-catchments of the watershed (Fig. 6), they are described separately in this section.

Table 7 Population and dwelling statistics for the Rocky View County and the City of Calgary (Statistics Canada 2012)

		Population			Private dwellings occupied by usual residents
Geographic name	CSD type	2006	2011	Change rate (%)	2011
Rocky View County	Municipal district	33,173	36,461	9.9	12,077
Calgary	City	988,812	1,096,833	10.9	423,417

Table 8 Parameters for the smart growth scenario

Change type	Number of changed cells	Minimum patch size	Number of changed patches
Change to under development	30,740	5,841	6
Change from under development to urban residential	23,180	5,845	5
Change from under development to country residential	7,560	6,042	1

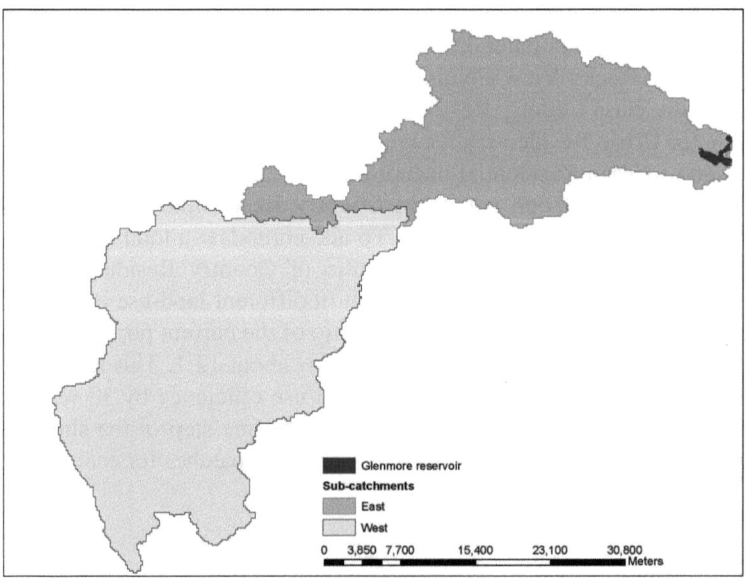

Fig. 6 East and west sub-catchments of the Elbow River watershed delineated using a digital elevation model

The east sub-catchment is dominated by built-up areas and agriculture. Due to the considerable growth of built-up areas (117 %) over the period 1992–2010, the evergreen and deciduous forest areas have been reduced by about 8 % and 11 %,

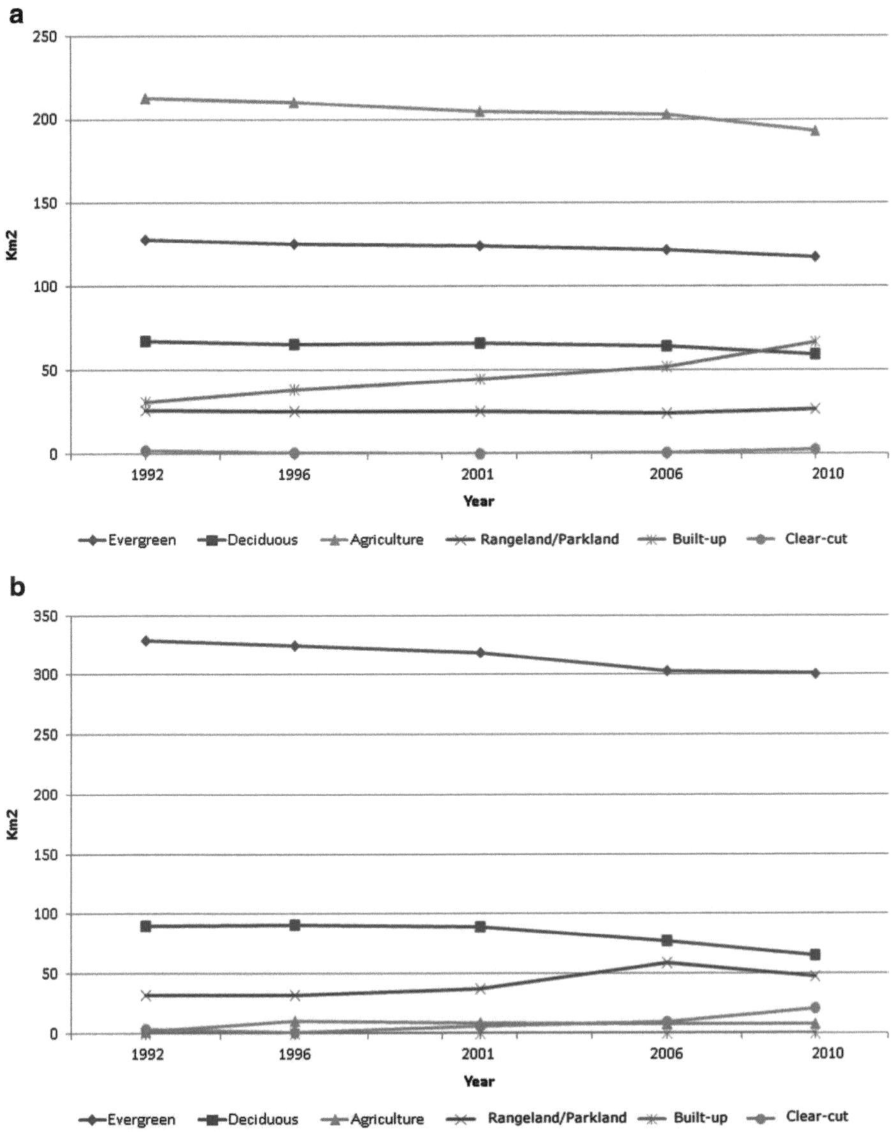

Fig. 7 Land-use changes for the period 1992–2010 in the east sub-catchment (**a**) and the west sub-catchment (**b**)

respectively along with agricultural areas (9 %) (Fig. 7a). Areas of rangeland/parkland have increased by 3 %.

The west sub-catchment is dominated by evergreen and deciduous forests. From 1992 to 2010, evergreen forest was reduced by 8 %, while the reduction is 28 % for the deciduous forests (Fig. 7b). Clear-cuts are minimal in 1992, but start increasing

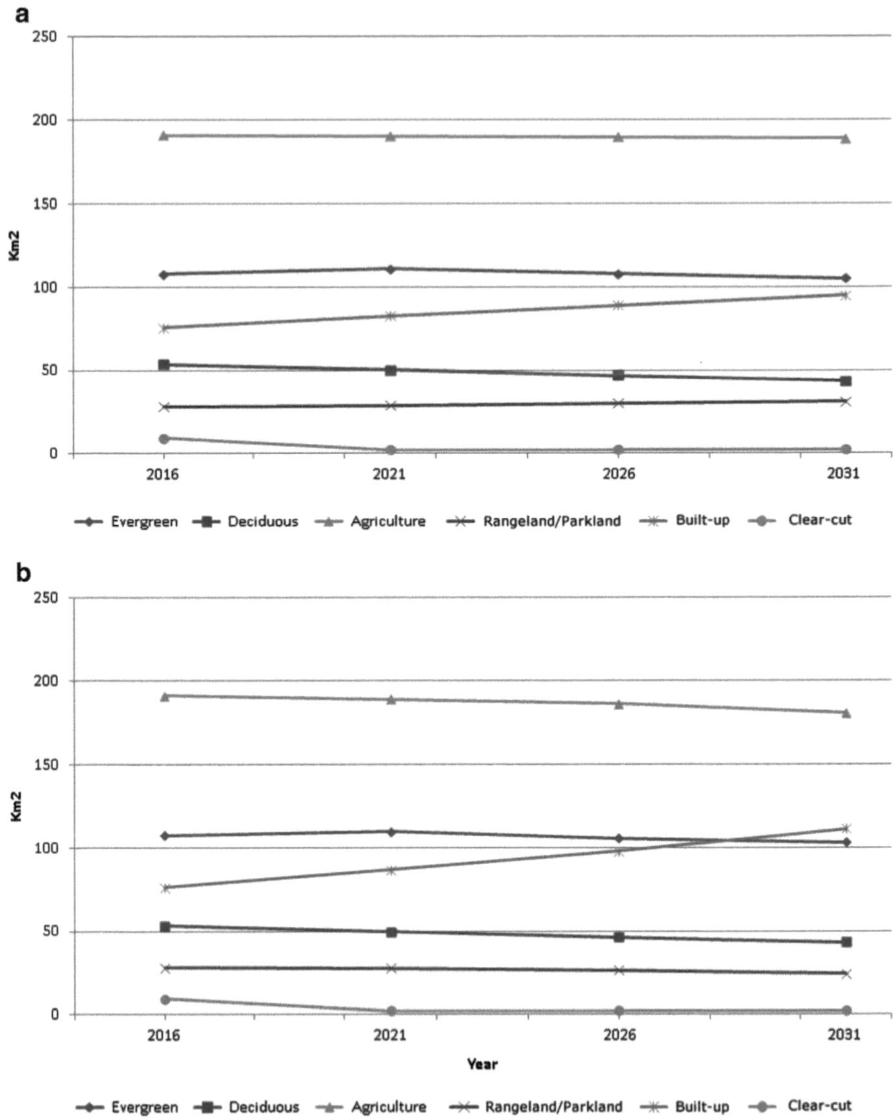

Fig. 8 Simulated land-use changes during the period 2016–2031 in the east sub-catchment of the Elbow River watershed based on scenarios BAU, RV-LUC, BC-LUC (**a**) and scenario P-LUC (**b**)

in the year 2000 to reach a peak value in 2010 (2.7 % of the sub-catchment). This results in an increase in rangeland/parkland.

Since the majority of the land-use changes appear within the east sub-catchment, the analysis of future land-use changes are only presented for this part of the watershed. The trends in the simulated land-use changes appeared to be the same for the scenarios BAU, RV-LUC and BC-LUC, and are presented in a unique graph in Fig. 8a. The growth of built-up areas reaches 25 % with a corresponding reduction of agriculture (1 %), evergreen (2.6 %) and deciduous (19 %) areas. For the scenario P-LUC (Fig. 8b), there is a substantial growth of built-up between 2016 and 2031 (46 %), while the areas of agriculture, evergreen and deciduous decrease by 5 %, 4 %, and 19 %, respectively. The higher change rate for built-up reflects the projected population growth represented in that scenario.

Despite the same rate of land-use change for the scenarios BAU, RV-LUC, BC-LUC, they have generated different spatial patterns due to the different spatial constraints applied during the simulations (Fig. 9). In the BAU scenario, new areas of built-up are sparsely distributed west of Calgary compared to the scenarios RV-LUC and BC-LUC, where concentrated built-up areas appear within the Rocky View County and in the area of Bragg Creek respectively. The scenario P-LUC generates more built-up areas appearing further west of the city of Calgary and in the north part of the watershed than the other scenarios. The spatial distribution of built-up areas in this scenario (P-LUC) is the same as for the scenario BAU.

3.2 Sub-Area of the Watershed

Figure 10 illustrates the land-use changes that occurred in the sub-area of the watershed adjacent to Calgary from 2003 to 2011. While 14 land-use classes were identified in the historical land-use maps, the transitions that happened between the five dynamic land-use classes are the most relevant, namely, from Agriculture and Forest to Under Development and from Under Development to Country Residential and Urban Residential. Under Development is a transitional class that is rapidly converted to the residential land-use classes. The most important changes happened between Under Development to Urban Residential. From 2003 to 2011, a total area of 3.65 km^2 of Urban Residential was created, representing an increase of 98 %. It can be observed that most changes to Urban Residential happened close to the existing Urban Residential and new Country Residential areas tend to appear at the fringe of existing Country Residential. Inside the city of Calgary's boundary, Country Residential (i.e., low density residential areas) can change to Under Development and then to Urban Residential.

The projected land-use changes under the three scenarios were analyzed for the simulation period of 2011–2041. The projected area for each land-use class is shown in Figs. 11, 12, 13, and 14.

For the class Country Residential, the growth trend is considerably different in the business-as-usual scenario than in the other two scenarios (Fig. 11). While it keeps expanding in a steady rate in the business-as-usual scenario (an average of 0.87 km^2 every 3 years), it varies slightly in the two other scenarios (an average of 0.16 km^2 for the protective growth scenario and 0.15 km^2 for the smart growth scenario), even decreasing after the year 2035. The reason is that in these two scenarios, the development of Country Residential is limited. At the same time,

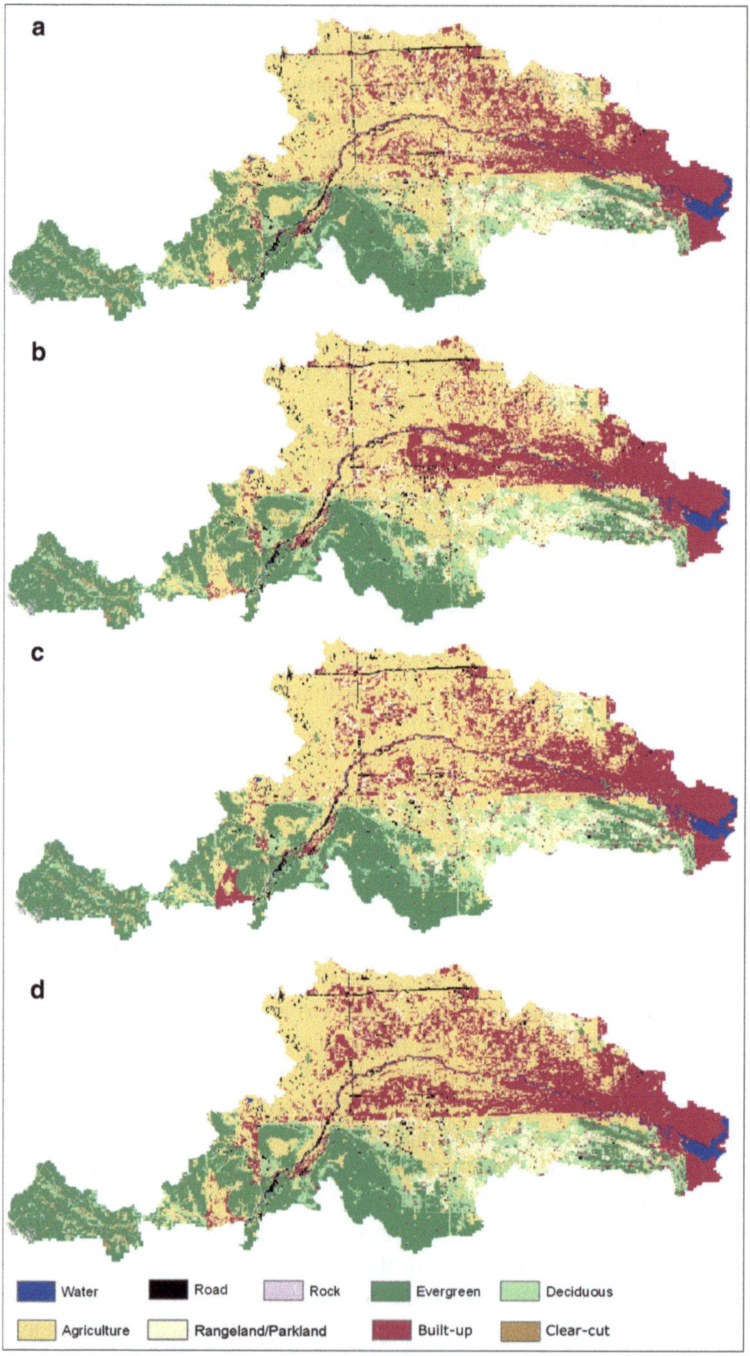

Fig. 9 Forecasted land-use maps for the year 2031 in the east sub-catchment of the Elbow River watershed according to the scenarios BAU (**a**), RV-LUC (**b**), BC-LUC (**c**), and P-LUC (**d**)

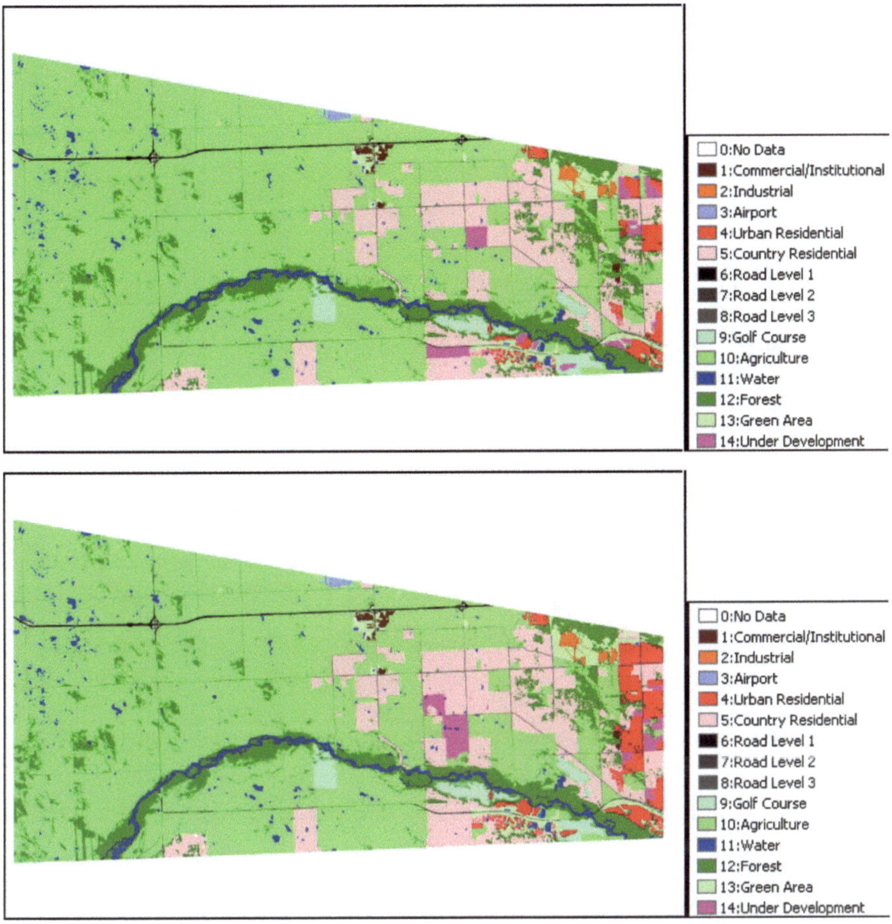

Fig. 10 Land-use changes that occurred from 2003 to 2011 in the eastern Elbow River watershed

large areas of Country Residential are converted to Under Development then to Urban Residential in the followings years. The net increase for Country Residential between 2011 and 2041 is 8.75, 1.64 and 1.51 km^2 for the three scenarios respectively, which indicates an increase of 30.3 %, 5.7 %, and 5.2 % for the entire period.

For the class Urban Residential (Fig. 12), a similar trend can be observed for the three scenarios. The area generated by the business-as-usual scenario (i.e., 11.24 km^2) is close to the one generated by the protective growth scenario (i.e., 12.34 km^2), while the area generated by the smart growth scenario is slightly smaller (i.e., 9.10 km^2). Represented as percentage of increase, it is 151 %, 166 %, and 122 % for the three scenarios, respectively.

The agriculture area (Fig. 13) declines in the three scenarios, but it is more pronounced in the business-as-usual scenario (an average of 1.63 km^2 every 3 years) and less important in the smart growth scenario (an average of 0.73 km^2

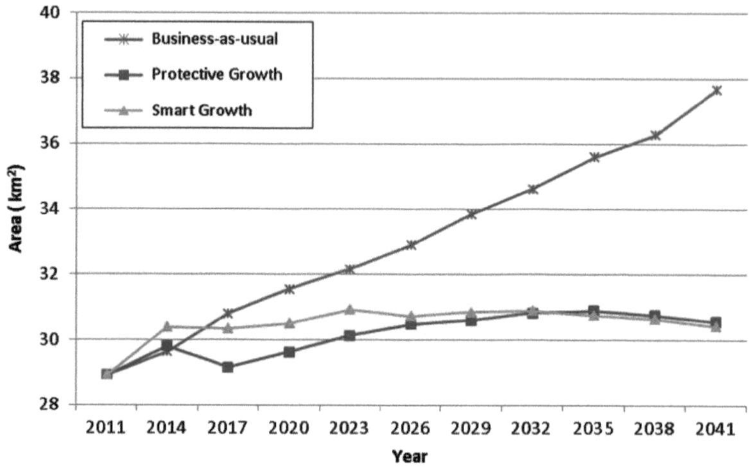

Fig. 11 Projected area for country residential over the simulation period

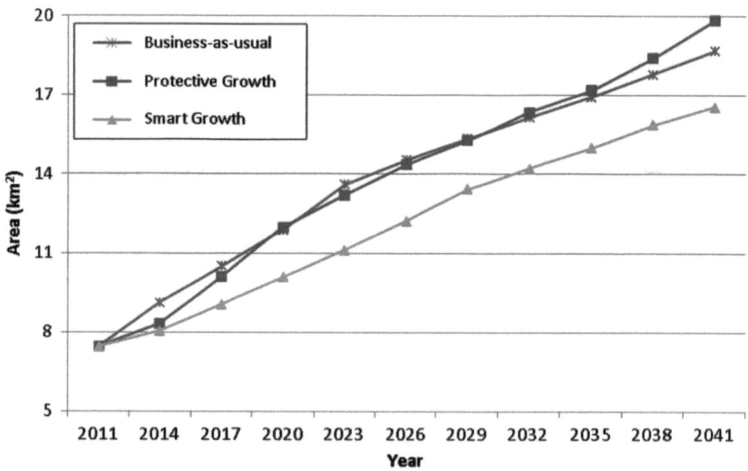

Fig. 12 Projected area for urban residential over the simulation period

every 3 years). The difference between the decreased areas becomes larger over time for the three scenarios. From 2011 to 2041, the total agriculture area dropped of 16.3, 11.0, and 7.3 km^2 for the three scenarios respectively, which corresponds to 10.5 %, 7.1 %, and 4.7 % of the total agriculture area in 2011.

The decrease of forest area (Fig. 14) is considerable in the business-as-usual scenario (i.e., 3.78 km^2 for the simulation period), while it is less pronounced in the two other scenarios (i.e., 1.71 km^2 in the protective growth scenario and 1.67 km^2 in the smart growth scenario). This suggests that the difference of the conversion from non-developed lands to developed lands between the protective growth and the

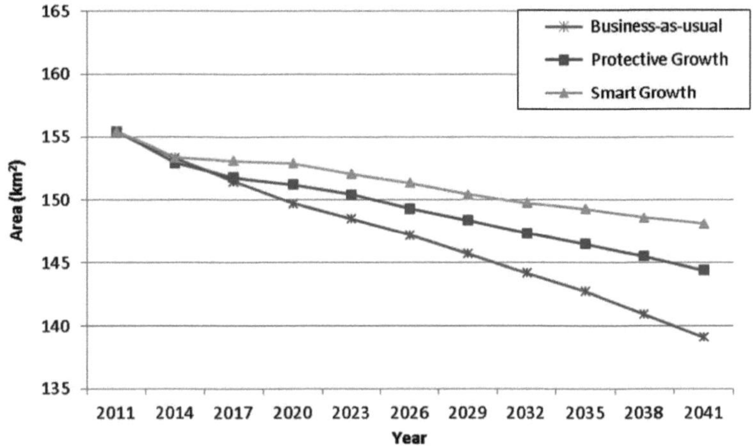

Fig. 13 Projected area for agriculture over the simulation period

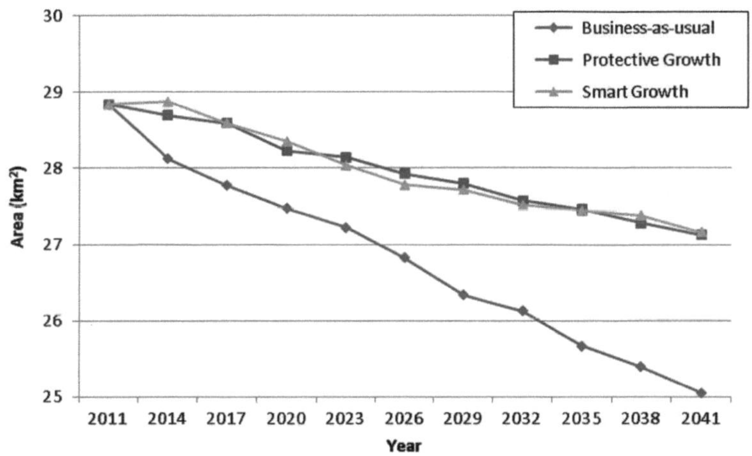

Fig. 14 Projected area for forest over the simulation period

smart growth scenarios is mainly generated by the amount of agriculture land consumption.

The projected land-use maps for the year 2041 under the three scenarios are displayed in Fig. 15. A first observation is that the land-use patterns produced by the CA model are compact and consistent with those observed in the historical maps.

The class Country Residential expands from east to west in the three scenarios. The expansion is more important in the business-as-usual scenario than in the two other scenarios, covering an area of 37.67 km^2 and resulting in an additional 7 km^2 of development (Table 9). In the business-as-usual and smart growth scenarios, the expansion of Country Residential tends to follow a space-filling pattern, where new

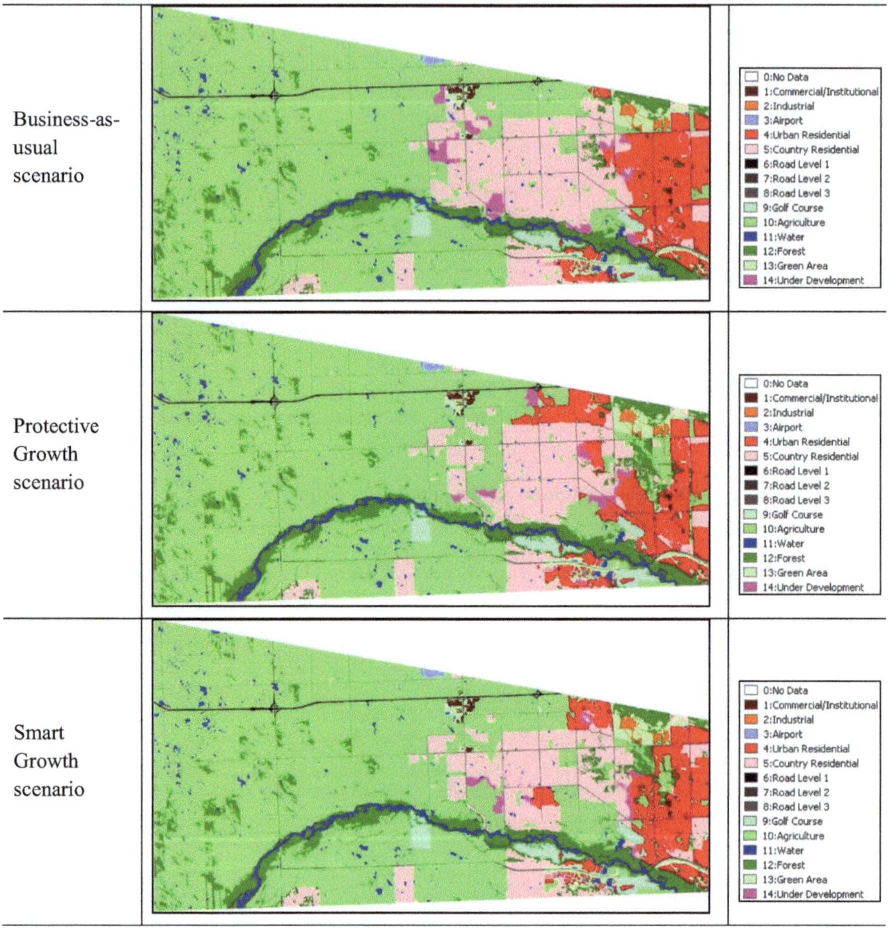

Fig. 15 Projected land-use maps for the year 2041 for the three scenarios

developments occur around the existing Country Residential areas. With the incorporation of agriculture and water constraints in the protective growth scenario, the development of Country Residential still appears around the existing Country Residential areas, but it avoids the good agriculture lands and lands that are too close to the river.

The expansion of the class Urban Residential occurs from east to west. In the business-as-usual scenario, almost all the non-developed lands within the city of Calgary boundary were converted to Urban Residential. In comparison, in the protective growth scenario, the good quality agriculture lands inside the Calgary limits were excluded and the growth of urban residential is therefore concentrated along the north transportation corridor (Highway 1). The smart growth scenario combines the trend of both the business-as-usual and the protective growth scenario, in which Urban Residential expands mostly from east to west, but is also

Table 9 Projected area of each land-use class under the three scenarios in 2041

Year	Business-as-usual scenario(km^2)	Protective growth scenario(km^2)	Smart growth scenario(km^2)
Country residential	37.67	30.56	30.43
Urban residential	18.69	19.83	16.55
Agriculture	139.13	144.42	148.13
Forest	25.05	27.13	27.16

concentrated along Highway 1. However the total area of Urban Residential generated in the smart growth scenario (i.e., 16.55 km^2) is noticeably smaller than the one generated in the business-as-usual scenario (i.e., 18.69 km^2) and the protective growth scenario (i.e., 19.83 km^2).

In both the business-as-usual and smart growth scenario, the agriculture land closed to the existing Country Residential and Urban Residential tends to be converted to urban land. While in the protective growth scenario, the good quality agriculture lands are kept untouched and the land development at proximity of the river is limited. The agriculture area in 2041 is 139.13, 144.42, and 148.13 km^2 for the three scenarios respectively. These preserved areas, although relatively small in size, add an important contribution to the ecosystem and agriculture in the study area.

A large portion of forest located at proximity of Urban Residential in the business-as-usual scenario was converted into Urban Residential, while it was preserved in the other two scenarios. By 2041, the protective growth and smart growth scenarios will preserve about 2 km^2 of forest comparing to the business-as-usual scenario. The preservation in the protective growth scenario basically occurs because these forested areas are close to the good quality agriculture lands, while the preservation in the smart growth scenario is mainly due to the small requirement in Urban Residential area for the simulation period. With the continuing urban growth in the following years, the preservation cannot be guaranteed if no further protection is involved.

An important objective for land-use planning and scenario testing is to keep the land-use system sustainable over long time period and limit the land consumption (i.e., agriculture and forest). As noted in Table 10, the area consumed is important for the year 2014 (i.e., 1.8 % for the business-as-usual scenario, 1.7 % for the protective growth scenario, and 1.3 % for the smart growth scenario) compared to the other years. While for the remaining of the years, the consumption rate tends to be more stable, corresponding to 1.2 % for the business-as-usual scenario, 0.7 % for the protective growth scenario, and 0.5 % for the smart growth scenario.

For the whole simulation period, as expected, the business-as-usual scenario is the least efficient in terms of land consumption with an area of 20.08 km^2 of non-developed land converted to developed land. Both the protective growth and smart growth scenarios consume less land (i.e., agriculture and forest) than the business-as-usual scenario. The protective growth scenario only consumes

Table 10 Land consumption for each scenario over the simulation period

	Business-as-usual scenario		Protective growth scenario		Smart growth scenario	
Year	Area (km^2)	Percentage (%)	Area (km^2)	Percentage (%)	Area (km^2)	Percentage (%)
2014	2.83	1.8	2.64	1.7	2.04	1.3
2017	2.18	1.4	1.25	0.8	0.53	0.3
2020	2.06	1.3	0.90	0.6	0.45	0.3
2023	1.45	0.9	0.88	0.6	1.15	0.7
2026	1.69	1.1	1.35	0.9	0.95	0.6
2029	1.98	1.3	1.07	0.7	0.99	0.6
2032	1.76	1.1	1.22	0.8	0.88	0.5
2035	1.90	1.2	0.98	0.6	0.57	0.4
2038	2.10	1.3	1.15	0.7	0.72	0.5
2041	2.13	1.4	1.26	0.8	0.70	0.4
Total	20.08	12.8	12.71	8.2	8.97	5.6

12.71 km^2 of non-developed land in order to accommodate the same population size as the business-as-usual scenario, which preserves 7.37 km^2 of non-developed land. By constraining the population and adopting higher land-use efficiency, the smart growth scenario only converted 8.97 km^2 of non-developed land from 2011 to 2041. The percentage of land consumption is 12.8 % for the business-as-usual scenario, 8.2 % for the protective growth scenario and 5.6 % for the smart growth scenario. Specifically, this later scenario consumes 36 % less land than the business-as-usual scenario, while the smart growth scenario only consumes about 44 % of land used in the business-as-usual scenario.

4 Conclusion

An analysis of the historical land-use maps for the period 1992–2010 at the scale of the watershed reveals two dominant trends: a considerable increase of urban areas in the east sub-catchment, at proximity of Calgary that occurred at the expense of forest and agriculture, and a decrease of forested areas in the west sub-catchment mainly due to industrial harvesting activities that intensified after 2000. When used to project population growth up to the year 2031 according to four scenarios in the east sub-catchment, the cell-based CA model generates different spatial patterns that reflect the spatial constraints of each scenario. In some scenarios, urbanization is concentrated around existing communities located west of Calgary, which facilitates the emergence of corridors of development along the main roads connecting these communities to Calgary. In other scenarios, urbanization occurs in a more diffuse way, driven by the existing road network, the distance to the Elbow River and to the Calgary center, and by the slope of the terrain, the view over the Foothills and the Rocky Mountains being an attractive factor of land development.

When focusing on the sub-area immediately adjacent to Calgary, the historical land-use maps generated from SPOT images at 5 m spatial resolution allow the discrimination of 14 land-use classes, including five dynamic classes: urban residential, country residential, agriculture, forest, and under development. Between 2003 and 2011, the most important land-use transition occurred from Under Development to Urban Residential; a total area of 3.65 km^2 of Urban Residential was created, representing an increase of 98 %. The three scenarios simulated with the patch-based CA model for the period 2011–2041 affect the main land-use classes differently. The BAU scenario generates a substantial increase of Country residential, mostly at the expense of agriculture and some forested areas, compared to a small increase in the two other scenarios. Urban residential increases in the three scenarios at a similar rate, while Agriculture decreases also at about the same rate. Forest decreases sharply in the BAU scenario and less considerably in the two other scenarios. Of the three scenarios, the smart growth maintains the largest agricultural area while maintaining the same forested area as the protective growth scenario. As expected, the BAU scenario consumes the highest percentage of land (12.8 %) compared to the protective growth (8.2 %) and smart growth (5.6 %) scenarios.

The CA model designed at the scale of the watershed provides useful information about alternative spatial distributions of urban areas that can occur according to spatial constraints imposed on land development. Of particular interest is the illustration that corridors of development along main roads might easily emerge when urbanization is concentrated in existing small communities located at about 30 km west of Calgary. The spatial distribution of the population in the watershed raises the issues of infrastructure, utilities, service accessibility, and water availability. Projecting land-use changes with a CA model can help decision makers find the most adequate distribution of population in regards to the resources that are available to sustain that population in the watershed.

In comparison, the CA model designed at fine spatial resolution offers several unique features. First, detailed land-use classes can be identified with a clear characterization of their boundaries. Changes of land-use internal structure reflecting the socio-economic attributes (e.g., country vs. urban residential) and ecological functions (i.e., green open space) can be detected. In addition, land-use relationships are easier to discern at fine scales, which is essential to implement local sustainable environmental policies and facilitate in situ decision makings (GLP 2005). As noted by Kok and Veldkamp (2001), these relationships are less apparent when the emergent land-use change patterns are observed at coarser scales where other processes, such as environmental or macro-economic factors, become dominant. Of particular interest in our study is how land consumption can be considerably diminished by encouraging the protection of sensitive areas and increasing the density of existing and new urban residential areas.

Land development at the periphery of a fast-growing city creates intense pressure on a wide range of resources. Understanding the land-use dynamics that occurs in these peri-urban areas is of primary importance for land-use planning. CA models designed at multiple scales can provide useful information on the dominant land-use transitions and their driving factors while allowing the exploration of alternative scenarios of development and a quantitative assessment of their impact on resources to ensure sustainability.

Acknowledgements This research has been funded by three research grants awarded to D. Marceau by GEOIDE, Tecterra, and NSERC Discovery and by an iCORE scholarship awarded to F. Wang from Alberta Innovates. We also thank Alberta Environment Sustainable Resource Development for their financial support and expertise.

References

Agterberg, F. P., & Bonharn-Carter, G. F. (1990). Deriving weights of evidence from geoscience contour maps for the prediction of discrete events, in Proceedings of the 22nd APCOM Symposium (Vol. 2, pp. 381–396), Berlin, September 1990: Technical University of Berlin.

Almedia, C. M., Batty, M., Monteiro, A. M. V., Camara, G., Soares-Filho, B. S., Cerqueira, G. C., & Pennachin, C. L. (2003). Stochastic cellular automata modeling of urban land use dynamics: Empirical development and estimation. *Computers, Environment and Urban Systems, 27*, 481–509.

Bonham-Carter, G. F., Agterberg, F. P., & Wright, D. F. (1989). Weights of evidence modelling: A new approach to mapping mineral potential. In F. P. Agterberg, G. F. Bonham-Carter (Eds.), *Statistical Applications in the Earth Sciences* (pp. 171–183). Geological Survey of Canada, Ottawa, Ontario.

BRBC (Bow River Basin Council). (2010). *Bow river basin: State of the watershed summary, Calgary, Alberta*, Bow River Basin Council. http://wsow.brbc.ab.ca/reports/BRBCWSOW BookletV2-Dec28.pdf, p. 23.

Calgary Economic Development, Demographics. (2010). http://www.calgaryeconomicdevelopment.com/liveWorkPlay/Live/demographics.cfm.

Canada Land Inventory. (2008). http://geogratis.cgdi.gc.ca/geogratis/en/option/select.do?id=CB6E057B-0D85-9B22-29DE-6351369A8B02

Central Springbank Area Structure Plan. (2001). *Department of Planning and Development, Municipal District of Rocky View No. 44*. http://www.rockyview.ca/LinkClick.aspx?fileticket=RKoWI2qipOo%3D&tabid=446&mid=858.

Chen, Z., Grasby, S. E., Osadetz, K. G., & Fesko, P. (2006). Historical climate and stream flow trends and future water demand analysis in the Calgary region, Canada. *Water Science & Technology, 53*(10), 1–11.

City of Calgary. (2009). *Population Picture*. Calgary: City of Calgary.

Civic Census Results City of Calgary. (2012). http://www.calgary.ca/CA/city-clerks/Documents/Election-and-information-services/Census2012/Final%202012%20Census%20Results%20book.pdf

Dickson, B. G., Prather, J. W., Xu, Y., Hampton, H. M., Aumack, E. N., & Sisk, T. D. (2006). Mapping the probability of large fire occurrence in northern Arizona, USA. *Landscape Ecology, 21*, 747–761.

Dietzel, C., & Clarke, K. C. (2007). Toward optimal calibration of the SLEUTH land use change model. *Transactions in GIS, 11*(1), 29–45.

ERWP. (2012). Elbow River Watershed Partnership. http://www.erwp.org/

Exelis Visual Information Solutions. (2012). http://www.exelisvis.com

Foley, J., DeFries, R., et al. (2005). Global consequences of land use. *Science, 309*, 570–574.

Fu, B., & Chen, L. (2000). Agricultural landscape spatial pattern analysis in the semi-arid hill area of the Loess Plateau, China. *Journal of Arid Environments, 44*, 291–303.

GLP. (2005). *Science plan and implementation strategy*. IGBP Report No. 53/IHDP Report No. 19, Stockholm, 64 pp.

Gutman, G., et al. (2004). *Land change science. Observing, monitoring and understanding trajectories of change on the earth's surface with remote sensing and digital image processing* (p. 459). Dordrecht: Kluwer Academic Publishers.

Hasbani, J.-G. (2008). *Semi-automated calibration of a cellular automata model to simulate land-use changes in the Calgary region*. Unpublished M.Sc. thesis, Department of Geomatics Engineering, University of Calgary, 96 pp.

Hasbani, J.-G., Wijesekara N., & Marceau, D. J. (2011). *An interactive method to dynamically create transition rules in a land-use Cellular Automata model, Cellular Automata – Simplicity Behind Complexity*. In Alejandro Salcido (Ed.), ISBN: 978-953-307-230-2, InTech. http://www.intechopen.com/books/cellular-automata-simplicity-behind-complexity/an-interactive-method-to-dynamically-create-transition-rules-in-a-land-use-cellular-automata-model

Imagine Calgary Plan for Long Range Urban Sustainability. (2006). http://www.imaginecalgary.ca/imagineCALGARY_long_range_plan.pdf

Kok, K., & Veldkamp, A. (2001). Evaluating impact of spatial scales on land use pattern analysis in Central America. *Agriculture, Ecosystems and Environment, 85*, 205–221.

Liu, X., Li, X., Shi, X., Zhang, X., & Chen, Y. (2010). Simulating land-use dynamics under planning policies by integrating artificial immune systems with cellular automata. *International Journal of Geographical Information Science, 24*(5), 783–802.

Parks Foundation Calgary. (2008). At home on the Elbow River. http://www.parksfdn.com/At%20Home%20on%20the%20Elbow%20River.pdf

Pawlak, Z. (1982). Rough sets. *International Journal of Parallel Programming, 11*(5), 341–356.

Plan It Calgary Workbook. (2007). http://www.calgary.ca/PDA/LUPP/Documents/Publications/plan-it-workbook-final.pdf?noredirect=1

Rindfuss, R. R., Walsh, S. J., et al. (2004). Developing a science of land change: Challenges and methodological issues. *Proceedings of the National Academy of Sciences, 101*(39), 13976–13981.

Rocky View County. (2012). Rocky View 2060 – Growth Management Strategy, Rocky View 2060. http://www.rockyview.ca/2060/reports/Final_GMS_Document.pdf

Rocky View County/City of Calgary Intermunicipal Development Plan. (2011). http://www.municipalink.ca/reports/docs/rocky_view_idp_june_2011.pdf

Statistics Canada. (2012). http://www12.statcan.gc.ca/census-recensement/2011/dp-pd/hlt-fst/pd-pl/Table-Tableau.cfm?LANG=Eng&T=302&PR=48&S=51&O=A&RPP=25

Trimble. (2012). http://www.ccognition.com.

Veldkamp, A. (2009). Investigating land dynamics: Future research perspectives. *Journal of Land Use Science, 4*(1–2), 5–14.

Verburg, P. H. (2006). Simulating feedbacks in land use and land cover change models. *Landscape Ecology, 21*(8), 1171–1183.

Wang, F., & Marceau, D. J. (2012). A patch-based cellular automaton for simulating land-use changes at fine spatial resolution. *Transactions in GIS*, in press. doi:10.1111/tgis.12009.

Wang, F., Hasbani, J.-G., Wang, X., & Marceau, D. J. (2011). Identifying dominant factors for the calibration of a land-use cellular automata model using rough set theory. *Computers, Environment and Urban Systems, 35*(2), 116–125.

Weins, J. A. (1976). Population responses to patchy environments. *Annual Review of Ecology and Systematics, 7*, 81–120.

Wijesekara, N., Gupta, A., Valeo, C., Hasbani, J. G., Qiao, Y., Delauney, P., & Marceau, D. J. (2012). Assessing the impact of future land-use changes on hydrological processes in the Elbow River watershed in southern Alberta, Canada. *Journal of Hydrology, 412–413*, 220–232.

The Spatial Morphology of Oporto's Urban Fringe

Miguel Serra and Paulo Pinho

Abstract We investigate the formation processes of suburban street networks, through the analysis of five study areas at Oporto's urban fringe, over a period of 55 years. We start by recreating their street grids on four different time periods through the common technique of map regression, extending it in order to make possible the identification of individual development operations (represented by their street layouts), occurring between sequential time periods within each study area. Space syntax is used to study the structural evolution of the complete street networks, while the individual morphologies of development operations are quantified and classified according to simple topological parameters. We observe different structural evolutions among the several study areas and also different frequencies for the morphological classes of development operations occurring therein. By crossing these two types of results, we show that the differences observed at the level of the entire networks may be explained by the also different morphologies of the individual developments constructing them through time. We conclude by suggesting that these findings offer some clues on how street networks could be planned from the bottom-up, by regulating street patterns at the very local level in order to achieve desired global outcomes.

1 Introduction

Worldwide, extensive metropolitan urbanization has produced new patterns of urban development, with physical and spatial characteristics that are utterly different from those of the historic, or traditional city (Levy 1999). Beyond their old urban cores, contemporary metropolitan regions extend far into the surrounding environment, assuming the peculiar forms of an 'urbanized landscape' or of a

M. Serra (✉) • P. Pinho
CITTA – Research Centre for Territory, Transports and Environment, Faculty of Engineering, University of Porto, Rua Dr. Roberto Frias, 4200-465 Porto, Portugal
e-mail: alvim.serra@fe.up.pt; pcpinho@fe.up.pt

D. Malkinson et al. (eds.), *Modeling of Land-Use and Ecological Dynamics*, Cities and Nature, DOI 10.1007/978-3-642-40199-2_5,

'landscaped city' – a mingling of urban spaces and built forms with natural and rural spaces, accompanied by a gradual disappearance of the traditional dichotomy between these two worlds (Sieverts 2003). From densely built, compact objects – easy to define by simple contrast with their surroundings – cities have now become porous and diffuse objects, sprawling across the landscape without clearly defined boundaries.

Even if no longer a novelty, metropolitan development was not followed by an objective conceptualization of its morphological characteristics. Instead, its novel non-orthodox morphology gave rise to a tide of theoretic criticism and repudiation, seeing in it the signs of a distressing and apparently unstoppable dismemberment of the city. The words of the influential American architectural critic, Lewis Mumford (Mumford 1961), provide a good example of such stance: "whilst the suburb served only a minority, it neither spoiled the country-side nor threatened the city. But now that the drift of the outer ring has become a mass movement, it tends to destroy the value of both environments without producing anything but a dreary substitute, *devoid of form* and even more devoid of the original suburban values" p. 506, our emphasis.

This theoretical state of denial, often more driven by aesthetical mistrust than by objective observation, has led to a general paucity of knowledge on the effective morphological nature of contemporary metropolitan and suburban environments. Indeed, the relevant question to ask is not if contemporary metropolitan form is aesthetically satisfactory or not, but rather what truly are its intrinsic morphological characteristics. As a recent paper (Prosperi et al. 2009) on the subject claims, "posing the concept of 'metropolitan form' as a question [...] is an absolute necessity at this stage of development of urbanized areas. [...] There is a persistent theme in the related literatures of architecture, urban design and urban and regional planning that the physical form of the contemporary metropolis is un-describable. [...] A new epistemology and a new language are needed for the question of metropolitan form" op. cit. pp. 1–2.

Yet, in a much older albeit brilliant paper, Friedman and Miller (Friedmann and Miller 1965) declare that in face of metropolitan contexts, "planners [...] are left in a quandary. Modern metropolitan trends have destroyed the traditional concept of urban structure, and there is no image to take its place. Yet none would question the need for such an image, if only to serve as the conceptual basis for organizing our strategies for urban development" op. cit. p. 313.

Forty-four years have passed between those two texts. Still, in view of their authors' claims, it seems that they were written just a few months apart. The truth is that further investigations on the subject of contemporary metropolitan and suburban morphologies are today more necessary than ever because, generally speaking, urban planning practice seems locked up in a kind of 'historical city syndrome': so bounded to the graspable and deeply interiorized image of the historical city (now obviously surpassed forever), that it seems unable to conceive any other type of urban morphological manifestation. In fact, as Michael Batty (Batty 2007) notes, this seems even to be a general condition: "if you were to ask the population at large to define a city, most would respond with an image much more akin to what a medieval or industrial city looked like than anything that resembles the urban world [...] in which we live" op. cit. p. 18. Indeed, cities were more or less alike for millennia (Mumford 1961); our experience of the contemporary urban vastness and fuzziness is still rather unripe. But there seems no doubt that the extended city is here to stay and that urban planning needs to understand this new type of urban form and to learn how to tackle it.

The work described in this chapter seeks to make a contribution towards a more objective understanding of contemporary metropolitan and suburban morphologies. We will report the results of an on-going research project, some of which were already described elsewhere (Serra and Pinho 2011), so here we will just stick to the most relevant findings. We have conducted a comparative morphological study between five representative suburban areas of Oporto's metropolitan region, observing the development of their street systems along 55 years. We try, as much as possible, to confine our work to quantitative morphological descriptions and analytical procedures, in order to avoid the above mentioned subjectivity and aesthetical bias of some discursive approaches to contemporary urban form.

We take street systems as our sole object of study, for two fundamental reasons. Firstly, because street systems have for long been recognized by urban morphology as the most structuring of urban form's three basic elements[1] (namely street, plot and building systems). This is because streets have the greatest temporal inertia, being highly resilient to subsequent changes after their construction (by contrast, buildings have the least temporal inertia, changing quite frequently); therefore, streets are also the most conditioning of all elements, in what regards subsequent transformations to themselves or to the elements that change more quickly (Levy 1999; Case-Scheer 2001; Kropf 2009, 2011; Whitehand 1992, 2001). Secondly, because the available analytical techniques to model and analyse the morphology of urban street systems (namely space syntax's techniques) are purely quantitative and have attained a significative level of empirical and theoretical support (Hillier 1989, 1996a; Hillier et al. 2005; Hillier and Vaughan 2007; Penn et al. 1998; Read and Bruyns 2007).

We address our study object at two different levels. At the global, or macro level, studying the temporal evolution of the structure of each area's street network as a whole; and at the local, or micro level, investigating the individual morphologies of the incremental development operations occurring within each area through time. At the macro level we use space syntax (Hillier 1996b; Hillier and Hanson 1984) as main research tool; at the micro level we use several simple morphological quantitative parameters, adopted (and adapted) from Stephen Marshall's (Marshall 2005) work, which have the advantage of being comparable to space syntax's global structural descriptions (both techniques will be detailed in the next section). These two levels of research and the crossing of their results, are intended to produce insights on how the individual morphologies of urban development

[1] See (Levy 1999) or (Kropf 2011) for details on the definitions and characteristics of urban form's basic elements.

operations give rise to global structures at the macro-level, through their progressive accumulation through time.

Contemporary metropolitan development is characterized by highly decentralized and uncoordinated growth processes, occurring in contexts of strong administrative and political fragmentation (EEA 2006). This is particularly true in the case under study. It seems difficult, given these factors and the territorial scope of metropolitan urbanization, to achieve substantial morphological control with top-down planning approaches. Instead, if we could find causal relationships between the individual morphologies that incrementally construct the city through time, and the global structural outcomes that they collectively produce at higher scales, the devising of bottom-up planning approaches to urban form would become possible (Marshall 2005, 2009). In other words, it would be possible to envision street pattern regulations for the very local level (which is liable of planning scrutiny and control), in order to achieve desired structural outcomes at the level of global street networks (which are difficult, if not impossible, to anticipate in detail today). As we will try to show, this work provides also some clues in that direction.

2 Methodology

In order to maximize the probability of observing significative morphological formation and transformation processes, we have selected as target areas the civil parishes[2] with highest urban growth rates over the last 50 years, belonging to the five municipalities that surround the city of Oporto. Because parishes' areas and their administrative limits are rather different, we considered a circular boundary of a 3 km radius for each study area (28.3 km^2, with centre on the centroid of the polygon formed by each parish's administrative limits). Figure 1 shows the location of these areas, as well as the limits of the chosen civil parishes and of the municipalities to which they belong. From now on, we will refer to these areas[3] as A, B, C, D and E, as shown in Fig. 1. These selection criteria produced some zones of overlap between study areas; but we accept this circumstance as natural, as these are excerpts of a larger and quite continuously urbanized zone. Therefore, the bordering context of each area (except case study E, the only to the south of the Douro river) is also part of the context of those which are nearer, because there are no abrupt urbanization discontinuities between them.

[2] The Portuguese system of territorial administration is divided, at the local level, in two tiers: the municipal level, and the civil parish level.

[3] The Portuguese names of the parishes are Custoias, Vermoim, Ermesinde, Rio Tinto and Mafamude, respectively.

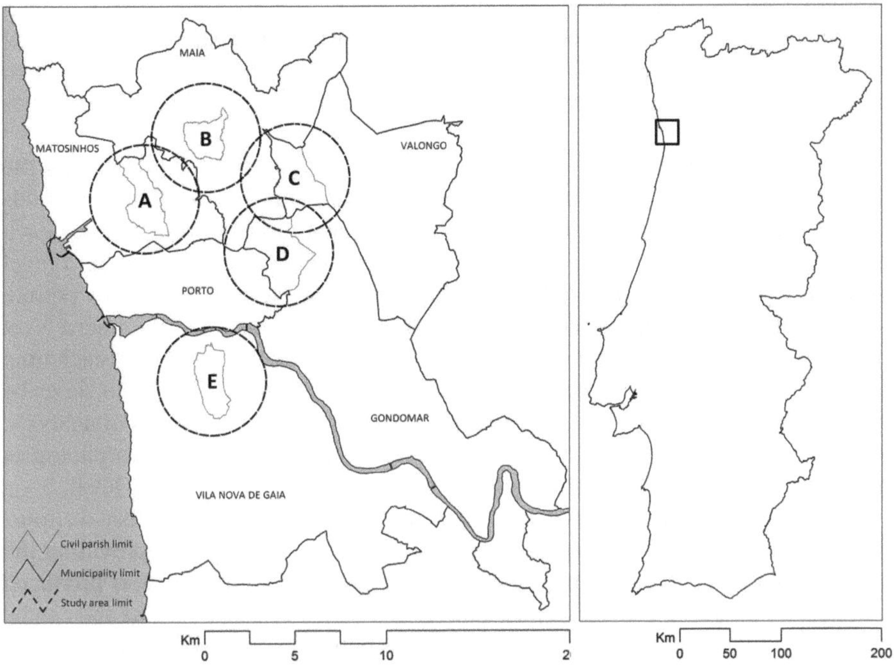

Fig. 1 Study areas location on Oporto's metropolitan region

Oporto's metropolitan region[4] is a dispersed, but continuously urbanized territory covering many cities and towns, most of them of ancient foundation. Unlike the typical metropolitan 'oil stain' pattern, produced by concentric growth from a dominant centre along a clear time-line, Oporto's metropolitan region has developed more by the densification of pre-existing and dispersed nuclei, with little differences in density or time of development. This polycentric urbanization pattern was fostered by a dispersed location of industry (starting late, well within the first half of the XX century) and thus also of employment, never generating the concentration of wealth and services at the central core, typical of early industrialized cities (Cardoso 2010).

However, prior to this industrial diffusion, a much older type of dispersed territorial occupation was already in place. The typical medieval land-division of the peninsular northwest, characterized by very small, family owned agricultural parcels, has generated a proliferation of small villages, hamlets and towns, and especially huge amounts of very old infrastructures (paths, roads, informal crossings), irrigating a former rural territory now thirsty for urbanization. The expansion of the built-up area has been supported in large part by this vast and labyrinthine

[4] The total region known as Greater Oporto Metropolitan Area (or GAMP, in its Portuguese acronym), spans 75 km of the Portuguese northern coast over an area of 1.885 km^2 and has a population of approximately 1.673.000 inhabitants.

capillary road network (mostly of rural origin) and by an important network of long and sinuous radial and radio-concentric roads. Over this old matrix, the creation of an arterial system of metropolitan highways has gained importance over the last two decades, deeply changing regional and local accessibility patterns (Domingues 2008).

Figure 2 summarizes the research methodology. The first step concerns the diachronic modelling of each study area's street system at several time periods and the identification of individual development operations occurring between each period (see below). This step provides the data to be explored at the macro level using space syntax (i.e. the study areas' entire street systems at each time period, subsequently modelled as axial maps); and those to be explored at the micro level (i.e. the sub-sets of individual development operations occurring between each time period), using some of the techniques proposed by (Marshall 2005) to describe individual street patterns. The crossing of the results of these two types of analysis, will then allow exploring potential causal relationships between the morphologies occurring at the micro level and the structures indentified at the macro level.

Each of these research steps uses different analytical procedures that we must first introduce in order to make our results comprehensible. Besides providing the raw material to be explored by macro and micro-analysis, the data sets produced during the diachronic modelling phase have their own analytical interest and can be explored using simple density and spatial distribution parameters. Space syntax's analytical procedures and theoretical background are central in our work and need also to be previously introduced, as well as the techniques proposed by Stephen Marshall (Marshall 2005) to quantify and classify individual street patterns. In the next three subsections we provide these methodological and theoretical reviews, which will constitute the context for the findings presented further ahead.

2.1 Diachronic Modelling of Street Systems in GIS

Both space syntax and the methods proposed by Marshall (Marshall 2005) are static analytical techniques: time is not a dimension of analysis. However, time and temporal dynamics are central objects of urban morphology and of this study in particular. Therefore, in order to make use of both these techniques but also to integrate time in our study, the street system of each study area had to be recreated at several historical moments. This was done by the common method of map regression (Kropf 2011; Pinho and Oliveira 2009) widely used in urban morphology, but implemented here in a Geographical Information System[5] (GIS) environment, making use of the possibility of storing data both in spatial and tabular formats.

[5] We used ESRI's ArcGIS 10.

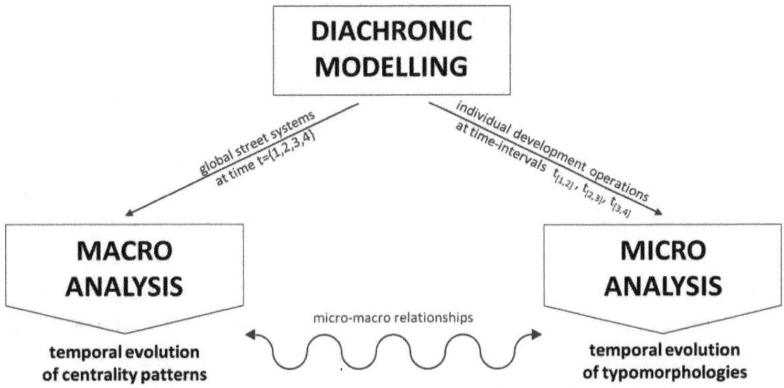

Fig. 2 Research methodology

We start by importing into GIS several digitized historical maps[6] and current vector street layers,[7] describing the street systems at four different points in time, $t = \{1,2,3,4\}$, corresponding respectively to the years[8] 1950, 1975, 1990 and 2005. We note the time intervals between these temporal moments as $t_{[1,\,2]}$, $t_{[2,\,3]}$ and $t_{[3,\,4]}$ (the first two intervals having a time-length of 25 years and the last of 15 years). The current vector layers (t = 4) are first used to georeference the digitized maps ($t = \{1,2,3\}$). Then, for each study area, we overlay the current vector layer over the most recent digitized map ($t = 3$) and tag all vector features that are not present on the digitized map. This is done through three binary attribute fields ([t3], [t2] and [t1], created on the vector layer's data table) describing the presence or absence of each vector feature at the periods $t = \{1,2,3\}$.

Initially, the [t3] field is entirely populated with 1 values; when a feature is identified as not present at $t = 3$, its value on the [t3] field is changed to 0; thus, recording its construction sometime during the time-interval $t_{[3,4]}$. We repeat this process until all features not present in the t = 3 map are indentified. Then, the values of the [t3] field are transposed to the [t2] field, the features where [t3] = 0 are excluded from the visualization, the digitized map corresponding to t = 2 is loaded and the process recommences (now changing the values of [t2] to 0, whenever a vector feature is not present in t = 2 map). The process is repeated for all time periods on all study areas. This method allows storing all information on the development of the street systems in a single vector layer for each study area, encoding the time of occurrence of each vector feature in tabular format instead of simply deleting them.

[6] Extracted from the Portuguese Military Map, covering the entire national territory at the 1:25.000 scale (edited by the Portuguese Military Geographic Institute, IGeoE).

[7] Extracted from urban digital cartographies, provided by the municipalities concerned in this study.

[8] The years 1950, 1975 and 1990 correspond to successive editions of the Portuguese Military Map; 2005 is the date of the municipal digital cartographies.

The extraction of the several temporal versions of each street system becomes a process of simple database querying. In the same way, we extract also the sub-sets of vector features created during each time interval.[9] The latter elements are subsequently visualized over current satellite imagery,[10] in order to further separate them into individual development operations. The large majority of these features occur as isolated clusters, usually corresponding to obviously different development operations. Whenever this is not the case (because several operations locate contiguously, or even intermingled), one can easily recognize their individuality through the visual inspection of the associated built structures, visible on the satellite imagery (e.g. the shape of buildings, type and colour of roofing, design style, pavements, etc). Besides providing the raw material for the macro and micro-analysis exercises, these data sets have their own analytical relevance. The simple visual inspection of the street systems at each $t = \{1,2,3,4\}$ (as depicted on the upper row of Fig. 3), is enough to infer some characteristic aspects of their evolution, by comparing their initial and final states. But we have also studied the two types of data sets depicted on Fig. 3 quantitatively, using basic density and spatial distribution parameters.

We compute *road density* by simply taking the total length (in meters) of the street system[11] of each study area at each $t = \{1,2,3,4\}$; because the area remains constant through time and is equal for each case study, total road-length becomes a direct and comparable measure of road density. We have also introduced a simple index expressing urban growth's[12] *spatial distribution*. This index is calculated by counting the number of individual urban development operations occurring within a circle with half the radius[13] of the study areas and centred therein, expressed as a percentage of the total number of development operations occurring at each time interval. Because the centres of the study areas are defined by the centroids of the civil parishes' with highest growth rates in the last 50 years; and because, in all cases, the main urban nuclei are today located at those positions; it is expectable that, under normal conditions, the majority of development operations would occur near the centre of each study area. Therefore, we consider urban growth to be concentrated when this index yields values above 50 %; when it yields values below 50 %, we consider it to be dispersed.

[9] For example, in order to extract the vector features created during $t_{[2,3]}$ we simply use the following SQL query condition: [t2] = 0 AND [t3] = 1.

[10] We used ESRI's 'World Imagery' layer, last updated in June 2013 and built-in ArcGIS 10. The layer features 0.3 m resolution imagery from DigitalGlobe, in the continental United States and parts of Western Europe (including Oporto's metropolitan region).

[11] We used the total length of axial lines in each street system after their conversion into axial maps (introduced further ahead), which is equivalent to total road-length.

[12] We take the previously individualized urban development operations as proxies of urban growth, more specifically of the growth of street systems.

[13] Thus, with 1.5 km radius and an area of 7.1 km^2.

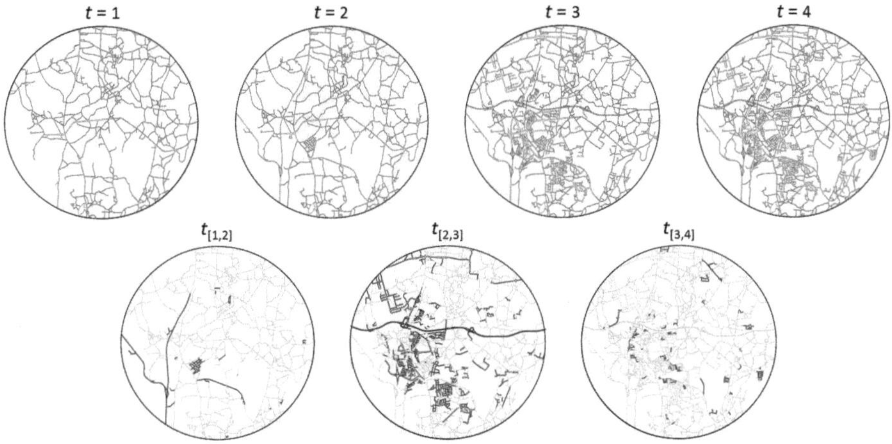

Fig. 3 Diachronic modelling of street systems in GIS (study area B). The street system at each t = {1,2,3,4} (*upper row*) and the street layouts of urban development operations occurring between each time period (*lower row*)

2.2 *Using Space Syntax to Study the Evolving Structure of Street Networks*

The study of urban spatial networks underwent major breakthroughs in the last three decades. Among other things, it has revealed that the topological constitution of such networks plays a decisive role in urban functioning. We now have the tools to begin understanding the reasons for cities being physically and spatially like they are, and why some parts of them are vibrant and seemingly fit for living in, while others are lethargic and apparently inappropriate. Space syntax (Hillier 1996b; Hillier and Hanson 1984) is the most fertile and operative manifestation of this approach, with direct applications on the urban planning and design fields as well as providing the theoretical framework for a growing research community.

Unlike other analytical approaches to urban form, syntactic analysis ignores the information contained in the built elements and addresses only the form of the open space system defined by them. It adopts several methods to represent the morphology of spatial systems as networks of elementary spatial units (or spatial primitives), easily translatable into graphs,[14] thus allowing for the quantitative analysis of their topological structure. One of such methods – axial mapping – is particularly useful for describing urban space. Here, we will concern ourselves only with this kind of spatial representation.

[14] A graph is a mathematical formalism, constituted by a set of nodes (or vertices) and a set of links (or edges), connecting those nodes. Graphs are used to represent any phenomenon in which there is a set of relations (the edges) between a number of objects (the vertices).

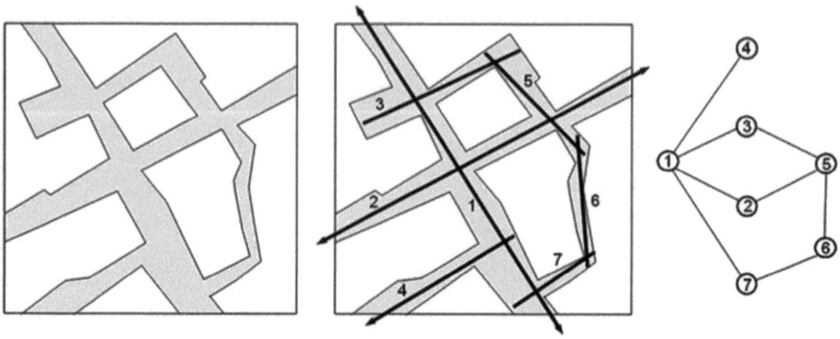

Fig. 4 A fragment of a hypothetical street system (*left*), its axial representation (*middle*) and the resulting axial graph (*right*). Numbers indicate axial lines and respective nodes in the axial graph

Axial maps aim at describing the spatial morphology of urban street systems by means of a very concise, skeletal representation. In an axial map, the convex geometry[15] of the planar representation of a given spatial system is approximately described by a set of interconnected straight lines (called axial lines) such that each line is extended as long as the geometry of the system allows, while ensuring that: i) any possible intersection with any other line is made; and ii) that the number of axial lines is minimal (thus, being also the longest lines meeting both conditions). Axial lines are translated into the nodes $V = \{1,\ldots,n\}$ of an undirected graph $G = (V,E)$, henceforth called axial graph, in which any pair of axial lines encoded as nodes, $i \in V$ and $j \in V$, are held to be adjacent, $i \sim j$, when they intersect on the axial map. The adjacency relations between all lines are encoded by edges $(i, j) \in E$, if and only if $i \sim j$. Figure 4 illustrates the process of axial representation and its subsequent graph encoding.

Graphs have long been used in geographical analysis, including the analysis of transportation networks (Garrison 1960; Hargett and Chorley 1969). However, the way street networks are represented as graphs in geography and transport research, differs significatively from space syntax's axial graphs. In the former type of representation, junctions are interpreted as nodes and streets as edges linking those nodes. This is called a *primal graph* representation (Porta et al. 2006a) and it suits well the study of transportation networks where nodes are significant points of terminus and interchange (as airports in airway networks, or junctions in large-scale road networks). However, this convention is less effective when the *morphology* of streets is the main focus of attention. But if a street system is first described by an axial map and axial lines are subsequently encoded as the nodes of a *dual graph* (Porta et al. 2006b), morphological specificities become apparent. Figure 5 shows the differences between the two types of graph representation.

[15] By convex geometry of a given polygonal shape we mean its decomposition into a minimal set of convex polygons (i.e. polygons in which any pair of points may be united by a line segment that remains entirely on the boundary or within the polygon itself). Circles, rectangles and triangles are convex polygons; however, any polygon with holes or dents is non-convex, because it has points that may only be united by segments that cross its boundary.

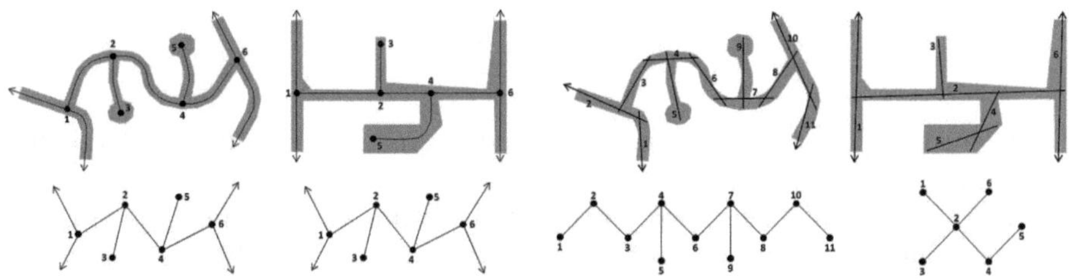

Fig. 5 Primal (*left*) and dual (*right*) graph representations of street networks

Although the two layouts show quite different morphologies, their primal graph representations (Fig. 5, left) have exactly the same topology (i.e. their graphs are homeomorphic). However, if both layouts are first represented as axial maps (Fig. 5, right), their different morphologies result in also different graphs. By using axial lines as nodes we are in fact encoding a morphological pattern into the graph, because axial lines are one-dimensional morphological descriptions of space (which are consistent insofar unobstructed sight and straight movement are possible along that line). Ultimately, what is being represented is the number of changes of visual axis, or the number of deviations from straight movement, that one must take to go from any point in the system to any other; both conditions are determined by the system's specific morphology. The correspondent axial graph is used to quantify the centrality (or accessibility) of each axial line in the system. Space syntax uses several graph measures for this purpose. We will only describe one of those, namely the measure of *integration*, which will be used for exploring the evolving centrality structures of our case studies.

Integration evaluates the relative proximity of each axial line to the overall spatial system. As a measure of proximity, it obviously deals with distances. However, the distance between two axial lines is of an unusual type (called topological distance, or *depth* within space syntax), measured in discrete units. If two axial lines are adjacent, the distance between them is 1 (independently of their actual metrical lengths); if they are not adjacent, the distance between them corresponds to the smaller number of edges separating their respective nodes in the axial graph (or geodesic distance). Therefore, the distance between two non-adjacent axial lines, can be seen as the fewest number of turns, or deviations from one visual axis to another, that one must take to move from one axial line to the other.

Integration is a normalized version of a simpler measure (which is actually equal to the *closeness centrality* of general network analysis), called *mean depth* within space syntax. In a connected and undirected axial graph $G = (V,E)$, the mean depth of a node $i \in V$ is given by the equation below, where k is the number of nodes in the graph and d_{ij} the length of each geodesic between i and any other node j (Hillier and Hanson 1984).

$$MD_i = \frac{1}{k-1}\sum d_{ij}$$

The mean depth of a node is highly dependent on the order of the graph (i.e. the number of its nodes, k) making impossible the comparison of values produced by different-sized graphs. This is only relevant in systems varying greatly in size and structure, but it constitutes a problem for the comparative analysis of cities and therefore a hindrance for research. In order to tackle this problem, mean depth values must be normalized according to some general 'yardstick'. Within space syntax, this is done by comparing the centrality of each node in a graph with k nodes, with the so-called D-value[16] for a special graph of the same order, or D_k, which is given by the following expression (Krüger 1989),

$$D_k = 2\frac{k\left[\left(\log_2 \frac{k+2}{3} - 1\right) + 1\right]}{(k-1)(k-2)}$$

Finally, the integration value of a node i (which is a standardized measure, comparable between graphs of different orders) is calculated by the equation below (Hillier and Hanson 1984) in which the numerator $(2MD_i - 2)/(k - 2)$ corresponds to MD_i rescaled[17] to vary only within the [0,1] interval.

$$integration\ (i) = \left(\frac{\frac{2MD_i - 2}{k-2}}{D_k}\right)^{-1}$$

The correspondence between the axial graph and the axial map allows centrality values to be represented also graphically. Figure 6 shows the global integration values of study area B at each time period, represented on the axial map according to a chromatic scale ranging from light-grey (low values) to black (high values). Such graphic representation is called 'integration pattern' and it makes visually evident the centrality structure of a given street network.

Like the generality of space syntax's graph measures, the integration value of each axial line may be calculated over the entire system (global integration) or over a restricted topological neighbourhood – or *radius* – around each line (local integration). For instance, for a given line, radius 1 includes all the lines directly connected to it; radius 2 includes also those lines, but still all other lines directly connected to them; and so on, regarding any desired topological radius around each line (Fig. 7). Thus, the centrality of each axial line may be assessed at different radii, or spatial scales, from the local level (e.g. the scale of a urban neighbourhood)

[16] The D-value of a node i in a graph with k nodes, corresponds to the mean depth of the 'root' node of a special type of graph with the same number of nodes (a so-called 'diamond graph'), in which depth values follow a strict normal distribution.

[17] This rescaled version of mean depth is called relative asymmetry (RA).

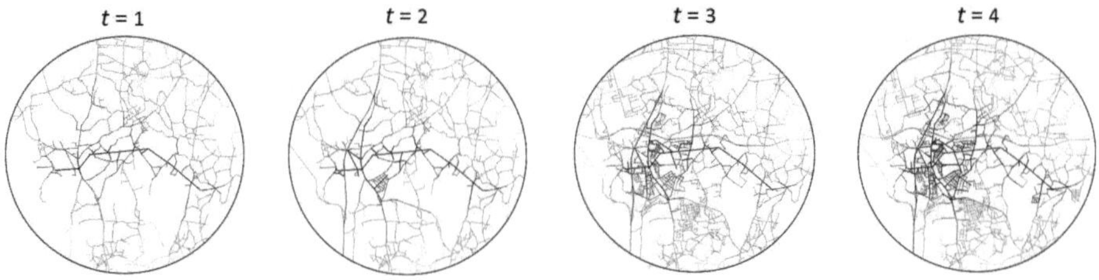

Fig. 6 Example of integration values represented on the axial map according to a chromatic scale (study area B)

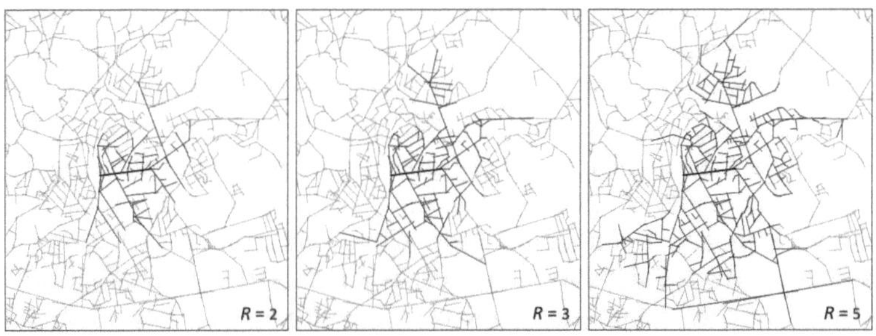

Fig. 7 Three radii (2, 3 and 5), or topological neighbourhoods (*black lines*), of an axial line of study area C (represented in *thick black*)

up to the scale of the entire spatial system under analysis (e.g. the scale of a city or even of an entire metropolitan region).

However, local and global centralities are not mutually exclusive: a given axial line may be central both at the local and global levels. This brings us to the last space syntax concept to be introduced, namely that of spatial *intelligibility*. For a given spatial network, this parameter is expressed by the correlation coefficient (R^2), between the values of local[18] and global integration (Hillier 1996b). When such correlation is high, it means that spaces that are locally integrated are also integrated at the global level, and that segregated spaces tend to be segregated at all scales. By contrast, when the correlation is low, it means that integration is more or less randomly distributed at the local and global scales, without any particular relation between locally and globally integrated (or segregated) spaces. Thus, this parameter can reveal deep centrality regularities in urban spatial networks, namely the presence (or absence) of a structural unity between the local parts and the wholes of those networks. As we will see in a moment, such structural unity can have significative impacts in the way we perceive and use urban spatial systems. In this work we studied both integration and intelligibility, finding clear differences

[18] Besides local integration, intelligibility is also quantified by the correlation between global integration and connectivity (i.e. the number of direct links each axial line has). Connectivity is the most local type of centrality (called degree centrality, in general network analysis).

between the several study areas and the way they have evolved through time. However, our findings may have relevance only at the light of the empirical validity of such properties; these are quickly reviewed in the following paragraphs.

The fundamental finding of space syntax, over which its theoretical *corpus* has been constructed, was the discovery of an intimate relationship between urban movement and spatial centrality. Movement (vehicular or pedestrian, from every origin to every destination) is the most basic and common use of urban space. However, urban movement does not distribute itself uniformly (or randomly) within urban street systems; quite the contrary, in every city one can always find bustling thoroughfares and also secluded streets. These inequalities in urban movement rates correlate strongly[19] with spatial centrality, as described by the above mentioned methods (Hillier et al. 2005; Penn et al. 1998). Because certain urban functions (like tertiary functions) thrive on the presence of urban movement (searching locations of high visibility and accessibility), while others are indifferent or even averse to high movement rates (like the residential function), it is a small step to conclude that urban functional patterns may also be explainable by urban spatial centrality patterns. And in fact they are, as several works (Chiaradia et al. 2009; Kim and Sohn 2002; Ortiz-Chao and Hillier 2007) have clearly demonstrated. Attracted to movement-rich locations, tertiary functions act then as additional movement destinations, creating a multiplying effect over the basic movement pattern induced by the street network itself (Hillier 1996a).

Spatial intelligibility, or the lack of it, has another type of structural significance. Firstly, from a purely morphological point of view, this parameter is capable of capturing the degree of structural order (or disorder) of a given spatial system (Hillier 1996b, 1999), if we understand 'order' in the sense of spatial linearity or axial continuity (which is not the same as pure geometric regularity). This type of spatial order seems to be closely related with some morphological regularities characteristic of naturally evolved urban street systems, as the seemingly universal Zipf distributions of axial line's lengths (Carvalho and Penn 2004), or the ubiquitous pattern of a foreground network made up of a few, quasi-linear main urban routes, set against a much more vast (and much less linear) background network of secondary streets (Hillier and Vaughan 2007; Hillier 1996b, 2002). Secondly, several studies have shown that, in unintelligible systems, the predictability of urban movement rates from spatial centrality decreases (Read 1999; Hillier et al. 1987, 1993; Park 2009). This seems to indicate that in such systems, movement (as an aggregated phenomenon) loses structure and becomes diffused, or with highly individualized patterns. And thirdly, other studies have also empirically shown that the property of intelligibility is indeed a relevant descriptor of a system's propensity for facilitating or for hindering spatial orientation (Conroy 2000; Haq 2003; Haq et al. 2005; Long et al. 2007; Tuncer 2007). For all these

[19] For example, (Penn et al. 1998) reports correlations of $R2 \sim 0.8$ for a very large sample of movement observations in London. Other studies have found similar correlations in many cities around the world.

reasons, syntactic intelligibility is also particularly relevant to this work, because suburban contemporary areas are usually qualified not only as 'labyrinthine' but also as 'structurally flawed' (at least when compared to the traditional city); therefore, this parameter may capture these characteristics.

2.3 Classifying and Quantifying the Morphology of Urban Development Operations

The last analysis technique was used to explore the morphology of urban growth at the micro level, as represented by the sets of individual development operations, identified before. We used some of the analytical tools proposed by Stephen Marshall (Marshall 2005) to classify and quantify these elements. Marshall's work is mainly concerned with two fundamental questions: how to classify street layouts in a consistent and systematic way and how to quantify their particular morphological characteristics. The analytical tools proposed by this author adopt a topological approach to street pattern classification, producing results that are consistent with those of space syntax.

Given the formal diversity of the development operations, the typomorphological classification criteria would have to be simple, yet sufficiently discriminant. We used two basic properties of street patterns mentioned in (Marshall 2005): the *internal structure* of street layouts; and the *external relationships* they establish with the surrounding grid. These two properties were assessed by inspecting the axial descriptions of the development operations (selecting the axial lines corresponding to their street layouts, extracted from each period's axial map), and how they were articulated with the existing street network (Fig. 8).

Internally, development operations were classified under two basic types, reflecting the presence or absence of urban blocks *within* their street layouts. In topological terms, these correspond respectively to axial sub-graphs with internal cycles[20] and to acyclic (or tree-like) sub-graphs. Throughout the first phase of this work (i.e. the recreation of the street systems at several temporal periods), we observed that the majority of new developments were small and simply composed by a few street segments, but that a minority were larger and commonly corresponded to layouts with internal blocks. As this is a basic difference between networks (i.e. cyclic or acyclic), moreover also indentified by (Marshall 2005) as a fundamental distinction between street layout morphologies, we classified the internal structure of development operations according to this criterion, defining

[20] In graph theory, a cycle is a closed path starting and ending at the same node and passing at least by two other nodes. In street networks, cycles correspond to urban blocks (or to islands of private space, completely surrounded by streets). A graph without cycles is called a 'tree' because, as trees, it has 'branches' but those branches never intersect.

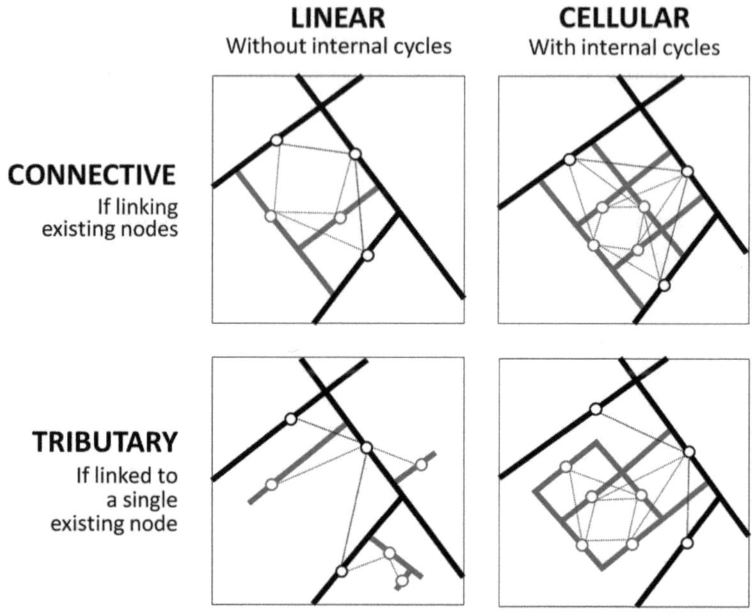

Fig. 8 The typomorphological classification system of the individual development operations

two morphological classes – cellular and linear – corresponding respectively to layouts with and without internal blocks.

We have also identified two recurrent types of external linkage, quite independent of the above mentioned internal properties. We have observed that some development operations had very few linkage points to the existing grid (usually just one, but sometimes two or three in the larger exemplars), thus not creating (or creating very few) connections between existing streets; conversely, other development operations (independently of their sizes) linked to several points of the existing grid, thus always creating new connections between existing streets (and also new potential circulation alternatives). In other words, the former type very rarely sub-divides existing spatial islands (or existing network cycles), while the latter type always creates new sub-divisions of existing spatial islands. These basic external connection possibilities are also mentioned in (Marshall 2005), where they are called respectively *connective* and *tributary*, designations that we have kept to characterize the external connectivities of the identified development operations. These types of internal structures and external linkages make a total of four possible typomorphological combinations (linear → connective/tributary, cellular → connective/tributary), summarized in Fig. 8. For each time interval, we classified as such each development operation and made an accounting of their relative quantities, as percentages of the total number of occurrences in each study area at each time interval.

The internal characteristics of each development operation (linear or cellular) are rather easy to establish. Externally, however, most developments are not so purely connective or tributary, although this criterion has shown to be effective if used with some degree of freedom.[21] But to cope with possible classification bias we used also two simple morphological parameters introduced in (Marshall 2005). Given the number of internal cells (graph cycles) and culs-de-sac (end-nodes of the graph, linked by cut-edges[22]) in a graph representation of a street network, it is possible to establish a 'cell ratio' and a 'cul ratio', as proportions of the total number of 'cells' and 'culs-de-sac'. Let C be the total number of cycles and D the total number of end-nodes in a street layout; then, the cell ratio (C_R) and the cul ratio (D_R) of that street layout will be given by,

$$C_R = \frac{C}{C + D} \quad and \quad D_R = \frac{D}{D + C}$$

In a pure tributary layout (a tree, without cycles) D_R will be one and the C_R will be zero. Conversely, in a pure cellular layout the D_R will be zero and the C_R will be one. More generally, in real-life mixed situations the values will vary, with the sum of both being one (Marshall 2005). We used these two basic quantifications, but for slightly different purposes. In an axial map, rings made by sequences of intersected lines correspond to cycles in the graph. In the same way, axial lines that do not lead to other lines (i.e. whose connectivity is 1), correspond to end-nodes. Thus, counting cycles and end-nodes in the axial representation becomes a rather simple task. However, we were not interested in quantifying the morphology of each development operation, but rather the morphological nature of urban growth at each time interval and in each study area; the idea was to obtain global morphological values, characterizing the extent to which urban growth was 'connective' or 'tributary'. These values would serve as morphological indicators *per se*, but also as benchmark values to control the typomorphological classification explained before. Thus, for the set of urban development operations of each time interval, we counted the total number of new cycles[23] and new end-nodes; with these values, we calculated the two ratios mentioned above, only now reflecting the global morphological composition of urban growth at each moment in time on each study area.

---

[21] For instance, some cellular developments have an extremely tributary character (and as such were classified), showing many cycles isolated from the surrounding network, even though they are externally connected with two existing streets and not just one. In the same way, although creating new links in the existing network, connective developments can come with some tributary appendices.

[22] In graph theory, a cut-edge (or bridge) is an edge which, if suppressed, divides the graph in two connected components.

[23] We did not count cycles of length 3, because on axial maps these are almost always trivial cycles, created by the simple intersection of axial lines in open space and not because there is anything built in the middle. Also, when counting new cycles, we took the care of discounting existing cycles; for instance, if a connective development operation divides a former single cycle into three cycles, the count of the new cycles will be two.

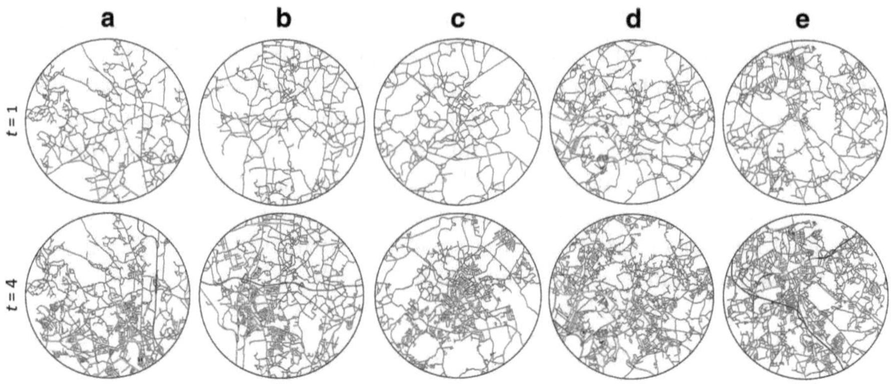

Fig. 9 The initial (t = 1, *upper row*) and final street systems (t = 4, *lower row*) of all study areas

3 Results

Our general result is in recognition of patterns of morphological change that are characteristic for each of the five study areas. This is despite all these areas being located within a 10 km radius from the centre of Oporto and considered as similar Oporto's suburbs. The street systems of t = 1 (before 1950), which are the initial spatial matrices over which urban growth will occur (Fig. 9, upper row), are visually describable as showing low street densities and a prevalent pattern of sinuous and narrow roads, with large undeveloped spatial islands, characteristic of local rural road networks (except study areas D and E,[24] which already show at t = 1 some small grid condensations of urban characteristics, i.e. smaller and more regular spatial islands and greater road density).

However, it is possible to visually discern on the final street systems differentiated morphological patterns (Fig. 9, lower row). Case studies B and C (and, to a certain degree, also A), evolved towards grids characterized by a dense, central urban core, surrounded by a sparser grid still with rural characteristics. In these cases, the distinction between urban and rural morphologies is rather clear. However, case studies D and E evolved into a kind of hybrid grid, with no clear spatial distinction between rural and urban components.

Even if all study areas show the same growth rates, their different final states may be explained (at least in part) by the evolution of urban growth's spatial distribution. The intensity rate of urban growth may be assessed by comparing the values of road density of each study area at all time periods (Fig. 10, left). Starting from different density levels (with density decreasing with increasing distance from the central city), all study areas show the same growth rate (i.e. their curves are approximately parallel), with a maximum intensity phase[25]

---

[24] These case studies are the nearest to the central city, and thus had an earlier urban development.

[25] Corresponding to the urban boom that Portuguese cities suffered between the 70' and the 90'.

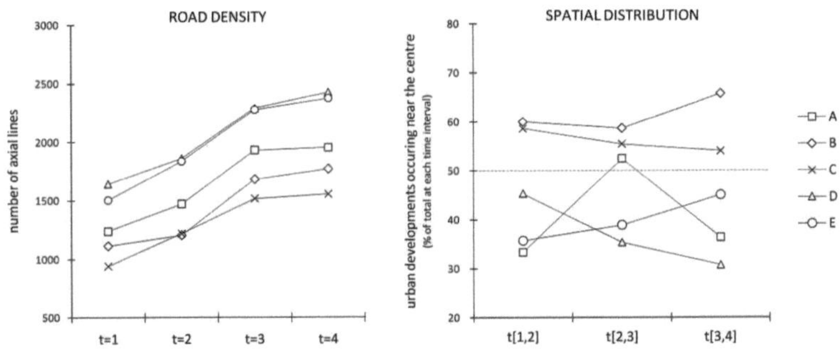

Fig. 10 Road density (*left*) and urban growth's spatial distribution (*right*), for all study areas at all time periods

during $t_{[2,\ 3]}$. Most of the transformations that the grids have undergone happen during this time interval. However, the spatial distribution of urban growth (measured according to the method described in Sect. 2.1) shows clear variations, both among study areas and across time (Fig. 10, right). Urban growth is always concentrated (>50 %) on study areas B and C, and always dispersed (<50 %) on study areas D and E. There are also periods of concentration and others of dispersion in the same study area, namely at A. However, during the time interval when urban growth is more intense ($t_{[2,3]}$), study areas A, B and C show a concentrated distribution, while D and E a dispersed distribution.

The space syntax results showed that differences were also present at the configurational level. Figure 11 (upper row) shows the integration patterns of all study areas at t = 1. Visual inspection of these patterns reveals few hierarchical differentiations between network spaces (i.e. patterns characterized by generally low integration values and very few dominant structures, with clear higher values or complex organization). This is general at all study areas except E, which shows a centre with some structure at this phase. But the other initial networks show just a few integrated long lines, being (for the most part) segments of old radial roads linking other towns to Oporto. These centrality patterns suggest that rural street systems do not favour any particular route structure, apart from a few more integrated spaces with distinct morphological characteristics (more linear). In contrast, urban spatial networks are in general characterized by highly heterogeneous centrality hierarchies.

The visualization of the final (t = 4) global integration patterns (Fig. 11, lower row) reveals different structural outcomes among the several study areas. Those in which concentrated growth was dominant (B, C and to a lesser extent also A), produced cohesive urban zones with clear centrality hierarchies. While in the cases were dispersed growth prevailed (D and E) we observe the stagnation, or even the dilution, of previous spatial hierarchies and a loss of structural differentiation between grid spaces. In spite of the intense grid transformations, and even if mean integration values kept increasing along time in all cases, in case studies D

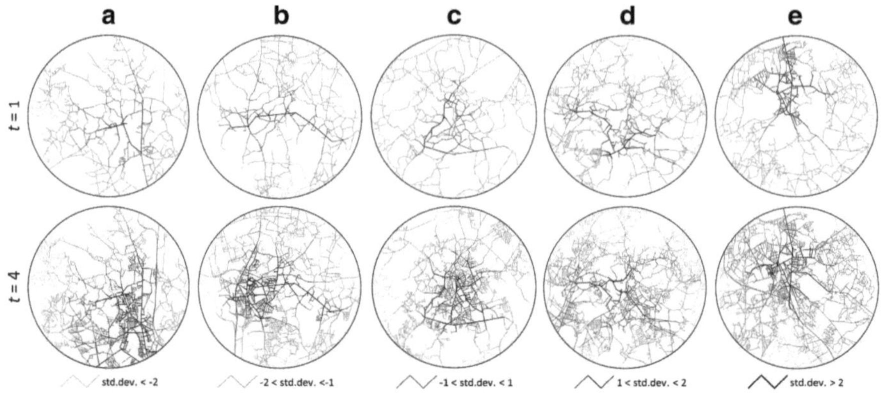

Fig. 11 The initial (t = 1, *upper row*) and final global integration patterns (t = 4, *lower row*) of all study areas

and E the dispersed growth pattern was not able to alter significantly the undifferentiated spatial character of the initial rural grids.

The evolution of the intelligibility values of the study areas (for both correlations of local integration and connectivity against global integration) showed an unexpected, yet quite relevant result, expressed in the charts of Fig. 12 (the yy axis represents R^2). A profound difference regarding the evolution of these parameters is noticeable, as case studies A, B, and C present ever increasing intelligibility values, while case studies D and E, show a sharp decrease during the period of most intense growth. The initial values are all low and close to each other, adding low intelligibility to the characteristics of rural street systems identified before. However, in case studies A, B and C, urban development manages to invert this situation and to raise intelligibility values considerably. However, in case studies D and E, in spite of an initial increase and of the similar urban growth rates, the values end up at the initial level.

Such a differentiated behaviour means that there are qualitative differences in the morphogenetic processes constructing the street systems throughout time. If suburban contemporary areas are marked by their labyrinthine character, these results show that some of the study areas diverged from that character, while others remained as unintelligible as they were initially, even if much more denser and developed. It is important to stress that these results refer to global structural properties of street networks emerging from urban growth, which is fundamentally a local process. In other words, such results describe emergent global effects, arising from discrete urban development events, operating at the local scale. The causes for such effects must therefore be sought at a lower level, that of the individual development operations constructing the networks over time.

Figure 13 (upper row) shows the frequencies of the typomorphological classes of urban development operations, as percentages of the total number of operations occurring during each time interval, for each study area. Again, the division in two

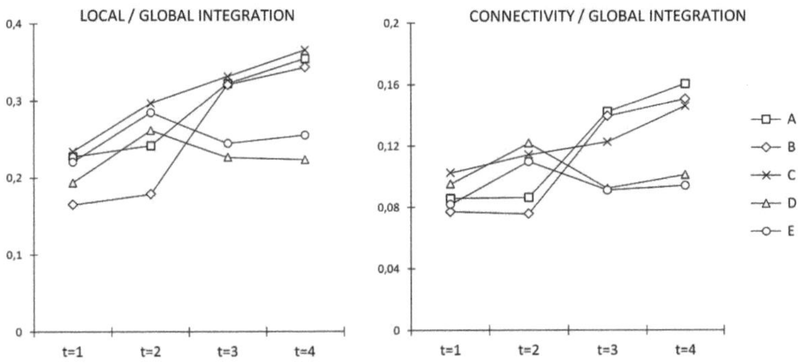

Fig. 12 Evolution of intelligibility values

groups is rather evident. In the first group (study areas A, B and C), connective typomorphologies are always prevalent over the tributary ones (both linear or cellular). However, in the second group (D and E) during $t_{[2,3]}$ (the time interval when growth is more intense) and regarding linear types (which are much more frequent than cellular types), tributary morphologies overcome connective ones. Thus, it is possible to say that the first group has a permanent connective grid construction; while in the second group, grid construction becomes predominantly tributary when most of the street system is produced.

The values of C_R ratio (Fig. 13, middle row), measuring the proportion of new cycles created at each time interval relatively to the creation of new end-nodes, is consentaneous with the previous result. At study areas A, B and C the creation of new cycles is always prevalent over the creation of new end-nodes (i.e. C_R values are always higher than 0.5); while at D and E, during the period of most intense growth, the creation of new cycles is surpassed by the creation of new end-nodes. These results show that the study areas of the first group were getting more cyclic (or more grid-like) along time; while the second group suffered a strong decrease in the creation of new cycles during $t_{[2,3]}$ (i.e. these areas got more acyclic, or more tree-like, during that time interval).

But in addition of being mutually supportive, these two results provide significant insights into the reasons behind the different evolution of intelligibility values, described previously. The charts in the lower row of Fig. 13 describe the variation of these values (Δ intelligibility) during each time interval (i.e. the intelligibility value of each time period subtracted by the value of the previous period). The relation between the former two parameters and intelligibility variation is rather evident: it is exactly during the period of most intense growth, when tributary forms overcome connective ones and new end-nodes surpass new cycles, that the intelligibility values of case studies D and E plunge. In other words, there seems to be a direct correspondence between the prevalence of tributary forms and intelligibility decrease. Indeed, this can be formally demonstrated by correlating the values of the C_R ratio at all time intervals and for all study areas, with the values of Δ intelligibility (Fig. 14). The correlation coefficients with both types of intelligibility

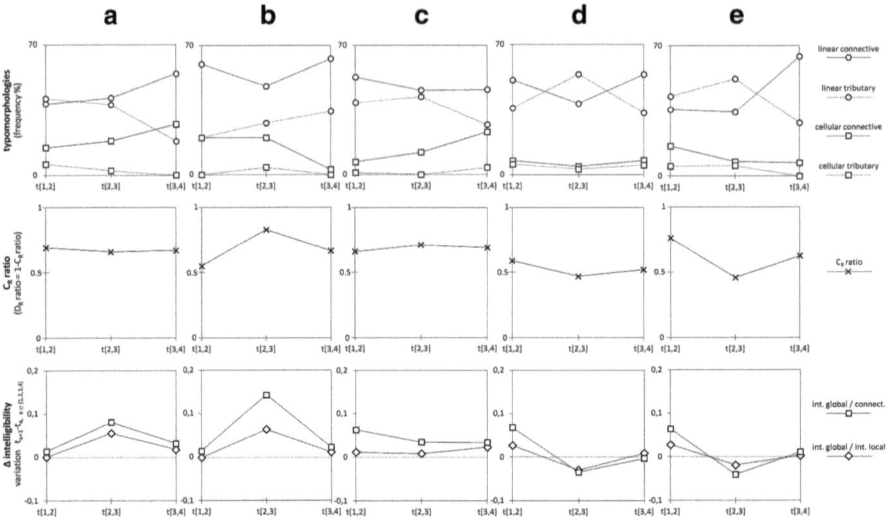

Fig. 13 The relation between frequencies of individual typomorphologies and intelligibility variation

quantification are quite significative ($R^2 = 0.55$ and $R^2 = 0.66$), showing that the more cyclic a street system gets the more intelligible it becomes. We only show the positive correlations with C_R, because C_R and D_R ratios are complementary (i.e. $C_R = 1 - D_R$) and thus the correlations with the D_R ratio are symmetric (i.e. same values, only negative).

It is important to stress that the new cycles taken into account in the calculation of both C_R and D_R ratios, are not just the internal cycles present only in cellular typomorphologies. New cycles created by the sub-division of previous ones (i.e. by the subdivision of existing spatial islands) are also accounted for and are much more frequent, being created both by linear and cellular typomorphologies (providing that they are externally connective). In fact, and besides the fact that they can also be tributary, cellular types are rather sporadic and could never explain the drastic variation of intelligibility values; what is determinant is not the distinction between cellular and linear internal structures, but between connective and tributary external linkages. What fosters the positive evolution of intelligibility, is the prevalence of urban development operations that sub-divide the existing grid and enhance its local connectivity; and what dictates its negative evolution, is the prevalence of urban developments that simply colonize existing streets to gain public access to their own layouts, but that do not create new connections between existing streets.

It seems unavoidable to conclude that the morphology of individual development operations, together with their spatial distributions (concentrated or dispersed), have been determinant to the evolution of the street networks' global structures. Such causal relationships are important; however, not because they are unlikely. In fact, with hindsight, one could say that they were expectable. Still, one

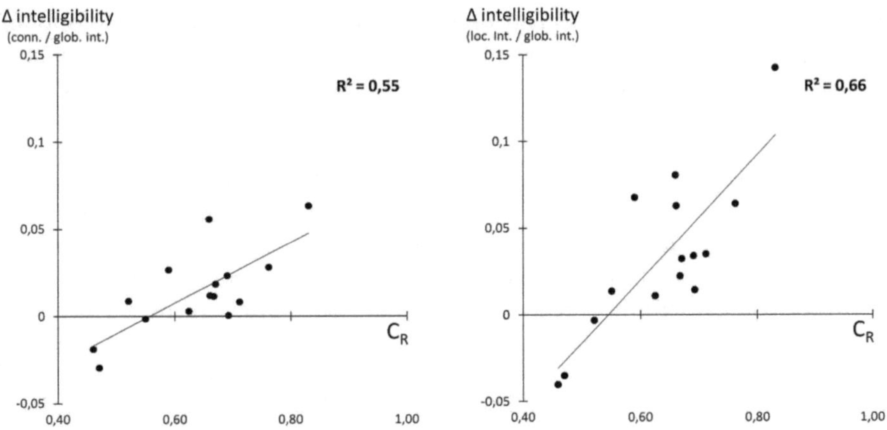

Fig. 14 Correlation between C_R ratio and the variation of intelligibility values

thing is to say that something is plausible; another is to show that such thing is factual and to measure its effects. Even if not claiming any generality for our results beyond the geographical context under study, we believe that they demonstrate that it possible to explore objectively contemporary suburban morphology and to understand how it arises from individual urban development events. Those seem to be nowadays the only possible targets for steering urban form. From the bottom up, in the same decentralized and uncoordinated way the contemporary city is produced, but with increasing knowledge on the aggregated consequences of individual morphological options.

4 Conclusions

Contemporary suburban areas present an analytic challenge to urban morphology. When approached at the level of their superficial characteristics they may seem a perversion of 'good city form', hopelessly disjointed and nonsensical. However, such morphological intractability seems to be avoidable by more quantitative approaches, capable of probing the deep structural characteristics of suburban street systems. The methodology adopted in this study was able to produce a different morphological picture of suburban development, not based on its odd superficial looks but rather on its underlying spatio-structural organization.

We provided evidence of different development patterns of suburban street systems, with also different structural outcomes. The study areas where urban growth was predominantly dispersed and led by non-connective typomorphologies, evolved into structurally flawed street networks characterized by low intelligibility values. In contrast, the study areas where urban growth was predominantly concentrated and composed by connective typomorphologies, evolved into better structured and intelligible street networks. Syntactic intelligibility is an indicator of the overall structural cohesion of a street network, something that accumulated evidence points as a relevant characteristic of well-functioning urban areas. It seems thus possible to say that our results provide also clues on how suburban street systems could be steered at the very local level, by creating morphological guidance and prescriptions that would lead to desired global structural outcomes.

Both growth concentration and street pattern prescriptions are parameters liable of being introduced in urban policies or planning instruments. The former aspect is already currently pursued in many situations. The latter not so much, because morphological prescriptions tend to focus more on the ephemeral elements of urban form (as buildings' architectural aspects) than on its more structuring and perennial elements (as street systems). However, the latter ought to be the main objects of morphological concern, because once laid out they will endure for long time lengths and will condition future urban form and functioning. Moreover, the establishment of basic morphological prescriptions for the street layouts of incremental urban development operations, seems quite feasible. Such prescriptions could be defined at the most basic morphological level, that of the internal and external connectivities of street layouts. This type of basic formulation would avoid the implementation difficulties that highly detailed or specific morphological prescriptions obviously entail. To promote a predominantly connective grid construction, shunning the tributary types, would not be difficult, both through planning instruments or through direct collaboration with agents of urban change. We would thus recommend that urban policies should add to their current concerns on urban dispersion, a further focus on street connectivity at the micro-scale.

References

Batty, M. (2007). *Cities and complexity: Understanding cities withh cellular automata, agent-based models and fractals*. Cambridge, MA: MIT Press.

Cardoso, R. (2010). *Space matters: Fine-tuning the variable geometry of cities*. CITTA 3rd Annual Conference on Planning Research, Porto.

Carvalho, R., & Penn, A. (2004). Scaling and universality in the micro-structure of urban space. *Physica A, 332*, 539–547.

Case-Scheer, B. (2001). The anatomy of sprawl. *Places, 14*(2), 28–37.

Chiaradia, A., Schwander, C., & Honeysett, D. (2009). Profiling land use location with space syntax: Angular choice and multi metric radii. In D. Koch, L. Marcus & J. Steen (Eds.), Proceedings of the 7th International Space Syntax Symposium. Stockholm: KTH.

Conroy, R. (2000). *Spatial navigation in immersive virtual environments, Unit for architectural studies*. London: UCL.

Domingues, A. (2008). Entensive urbanisation: A new scale for planning. CITTA 1st Annual Conference on Planning Research, Porto.

EEA. (2006). *Urban sprawl in Europe: The ignored challenge, EEA report*. Copenhagen: European Environment Agency.

Friedmann, J., & Miller, J. (1965). The urban field. *Journal of the American Planning Association, 31*(4), 312–320.

Garrison, W. L. (1960). Connectivity of the interstate highway system. Papers and Proceedings, Regional Science Association, Vol. 6 (pp. 121–137).
Haq, S. (2003). Investigating the syntax line: Configurational properties and cognitive correlates. *Environment and Planning B: Planning and Design, 30*, 841–863.
Haq, S., Hill, G., & Pramanik, A. (2005). Comparison of configurational, wayfinding and cognitive correlates in real and virtual settings. 5th International Space Syntax Symposium, Delft.
Hargett, P., & Chorley, J. C. (1969). *Network analysis in geography*. London: Butler & Tanner.
Hillier, B. (1989). The architecture of the urban object. *Ekistics, 56*(334/33), 5–21.
Hillier, B. (1996a). Cities as movement economies. *Urban Design International, 1*(1), 41–60.
Hillier, B. (1996b). *Space is the machine: A configurational theory of architecture*. Cambridge: Cambridge University Press.
Hillier, B. (1999). The hidden geometry of deformed grids: Or, why space syntax works, when it looks as though it shouldn't. *Environment & Planning B: Planning and Design, 26*(2), 169–191.
Hillier, B. (2002). A theory of the city as object: Or, how spatial laws mediate the social construction of urban space. *Urban Design International, 7*, 153–179.
Hillier, B., & Hanson, J. (1984). *The social logic of space*. Cambridge: Cambridge University Press.
Hillier, B., & Iida, S. (2005). Network effects and psychological effects: A theory of urban movement. COSIT 2005 – International Conference on Spatial Information Theory. Elliotville.
Hillier, B., & Vaughan, L. (2007). The city as one thing. *Progress in Planning, 67*, 3.
Hillier, B., et al. (1987). Creating life: Or, does architecture determine anything? *Architecture and Behaviour, 3*(3), 233–250.
Hillier, B., et al. (1993). Natural movement: Or, configuration and attraction in urban pedestrian movement. *Environment & Planning B: Planning and Design, 20*, 29–66.
Kim, H. K., & Sohn, D. W. (2002). An analysis of the relationship between land use density of office buildings and urban street configuration: Case studies of two areas in Seoul by space syntax analysis. *Cities, 19*(6), 409–418.
Kropf, K. (2009). Aspects of urban form. *Urban Morphology, 13*(2), 105–120.
Kropf, K. (2011). Morphological investigations: Cutting into the substance of urban form. *Built Environment, 37*(4), 393–408.
Krüger, M. (1989). *On node and axial grid maps: Distance measures and related topics*. London: Bartlett School of Architecture and Planning, UCL.
Levy, A. (1999). Urban morphology and the problem of moder urban fabric: Some questions for research. *Urban Morphology, 3*(2), 79–85.
Long, Y., Baran, P., & Moore, R. (2007). The role of space syntax in spatial cognition: Evidence from urban China. 6th International Space Syntax Symposium, Istanbul.
Marshall, S. (2005). *Streets and patterns*. New York/London: Spon Press.
Marshall, S. (2009). *Cities, design and evolution*. London/New York: Routledge.
Mumford, L. (1961). *The city in history: Its origins, its transformations and its prospects*. San Diego: Houghton Mifflin Harcourt Trade & Reference Publishing.
Ortiz-Chao, C., & Hillier, B. (2007). In search of patterns of land-use in Mexico City using logistic regression at the plot level. 6th International Space Syntax Symposium, Istambul.
Park, H.-T. (2009). Boundary effects on the intelligibility and predictability of spatial systems. 7th International Space Syntax Symposium, Stockholm.
Penn, A., et al. (1998). Configurational modelling of urban movement networks. *Environment & Planning B: Planning and Design, 25*, 59–85.
Pinho, P., & Oliveira, V. (2009). Cartographic analysis in urban morphology. *Environment & Planning B: Planning and Design, 36*(1), 107–127.
Porta, S., Crucitti, P., & Latora, V. (2006a). The network analysis of urban streets: A primal approach. *Environment & Planning B: Planning and Design, 33*(5), 705–725.
Porta, S., Crucitti, P., & Latora, V. (2006b). The network analysis of urban streets: A dual approach. *Physica A, 396*, 853–866.

Prosperi, D., Moudon, A. V., & Claessens, F. (2009). The question of metropolitan form: An introduction. *Footprint, 5*(Autumn), 1–4.
Read, S. (1999). Space syntax and the Dutch city. *Environment & Planning B: Planning and Design, 26*(2), 251–264.
Read, S., & Bruyns, G. (2007). The form of a metropolitan territory: The case of Amsterdam and its periphery. 7th International Space Syntax Symposium, Istambul.
Serra, M., & Pinho, P. (2011). Dynamics of periurban spatial structures: Investigating differentiated patterns of change on Oporto's urban fringe. *Environment & Planning B: Planning and Design, 38*(2), 359–382.
Sieverts, T. (2003). *Cities without cities: An interpretation of the Zwischenstadt*. New York: Spon Press.
Tuncer, E. (2007). Perception and intelligibility in the context of spatial syntax and spatial cognition: Reading an unfamiliar place out of cognitive maps. 6th International Space Syntax Symposium, Istanbul.
Whitehand, J. W. R. (1992). Recent advances in urban morphology. *Urban Studies, 29*(3–4), 619–636.
Whitehand, J. W. R. (2001). British urban morphology: The Conzenian tradition. *Urban Morphology, 5*(2), 103–109.

Vascular Plant Species Richness Patterns in Urban Environments: Case Studies from Hannover, Germany and Haifa, Israel

Sarah Matthies, Daniella Kopel, Stefan Rüter, Marina Toger, Rüdiger Prasse, Daniel Czamanski, and Dan Malkinson

Abstract The continuous expansion and growth of urban and settled areas result in a mosaic of open spaces which provide important habitat for species. Species richness within the urban matrix has been commonly studied in relation to urban-rural gradients, where the richness in open-space patches has been evaluated with respect to their location along this gradient. In this study we propose that additional factors may drive richness properties, namely patch size. We established a comparative study, where species richness patterns were compared between Haifa and Hannover, with respect to two driving factors: open-area patch size and its distance from the urban edge. These relationships were assessed for the overall number of vascular plant species and for native species only. Patches were identified by classifying aerial photographs of the cities, and surveying 32 patches in Hannover and 37 patches in Haifa which were randomly selected from the delineated patches. Results indicate that in both cities distance from the urban edge was not a significant factor explaining either the total vascular plant richness in the patches, or the native species richness. In contrast, both classes of species richness were significantly related with patch size. R^2 values for total richness were 53 % in Hannover and 45 % in Haifa. With respect to native species richness, patch size explained a higher proportion of the variance in Hannover where $R^2 = 73$ %, and a lower

S. Matthies (✉) • S. Rüter • R. Prasse
Institute of Environmental Planning, Leibniz Universität Hannover, Herrenhäuser Str. 2, D – 30419 Hannover, Germany
e-mail: matthies@umwelt.uni-hannover.de; rueter@umwelt.uni-hannover.de; prasse@umwelt.uni-hannover.de

D. Kopel • D. Malkinson
Department of Geography and Environmental Studies and The Golan Research Institute, University of Haifa, Haifa, Israel
e-mail: dkopel@kopel-ranch.com; dmalk@geo.haifa.ac.il

M. Toger • D. Czamanski
Faculty of Architecture and Town Planning, Technion – Israel Institute of Technology, Haifa, Israel
e-mail: marinat@technion.ac.il; danny@czamanski.com

D. Malkinson et al. (eds.), *Modeling of Land-Use and Ecological Dynamics*, Cities and Nature, DOI 10.1007/978-3-642-40199-2_6,

proportion of the variance in Haifa (R^2 = 33 %). These preliminary results indicate strikingly similar driving factors in two urban landscapes which are characterized by fundamentally different histories and environments.

1 Background

Ever since mankind developed permanent settlements, biodiversity patterns of the occupied areas and their surroundings have been modified by human activity. Soil sealing, drainage, and waste production are modern examples of means by which human activities may change environmental conditions (e.g. water supply, nutrient supply, temperature) and, therefore, assemblages of species and structural elements (Ricklefs and Miller 2000). These changes are accompanied by an emergence of spatial patterns of different patches within urban areas. These patterns reflect the type and intensity of human activities in the urban matrix, like housing, business, or recreation. However, the mosaic of different patches created and maintained by humans form a mosaic of habitat types ranging from fairly natural to highly modified ones (Sukopp et al. 1995). Within cities, open spaces (e.g. parks, allotments, urban forests) in particular provide habitat to numerous species of different taxa (Knapp et al. 2008; Kühn et al. 2004; Meffert and Dziock 2012; Melles et al. 2003). They may even fulfill basic nature conservation functions as endangered or rare species colonize urban open spaces (Niemelä 1999). However, little is known about patterns of species richness and key factors that influence total and native species numbers as well as the relation between native and non-native species.

Many factors influence species richness in urban open spaces, like the heterogeneity of vegetation structures (Blair 1996), human disturbance (Guntenspergen and Levenson 1997), age of habitats (Drayton and Primack 1996), and surrounding land uses (Guntenspergen and Levenson 1997). Nevertheless, patch properties like the size of urban open spaces may be of prime importance for development and maintenance of species richness (Mörtberg 2001; Knapp et al. 2008; Meffert and Dziock 2012; Bolger et al. 1997).

In recent years an increasing number of studies attempted to explore the relationship between patterns of the urban matrix and biodiversity. Many of these studies describe the relationship as a linear gradient of decreasing urbanization from city center to fringe areas which is associated with a general trend of increasing species richness (e. g. Guntenspergen and Levenson 1997; Mörtberg 2001; Niemelä 1999). McKinney (2008), who reviewed 105 studies focusing on flora and fauna diversity along urban-rural gradients, points out that diversity was lowest in the most intensively urbanized areas, commonly located at the central districts of the cities. Approximately 70 % of these studies indicated that vertebrate and invertebrate diversity peaked in the least urbanized areas, whereas a similar proportion of the studies found vegetation diversity to be highest in moderate levels of urbanization. As increasing urbanization is correlated with increasing fragmentation (Wickham et al. 2000), the low species diversity in the central districts may be explained by a smaller size and increased isolation of open spaces as compared to areas with a lower degree of urbanization. Populations of such patches may not be viable on the long term as colonization probabilities may be reduced and/or the area may be too small to support a minimal viable population (Snep et al. 2006). The studies reviewed by McKinney (2008), however, assume that the degree of urbanization is positively correlated to the distance to the urban center.

Ramalho and Hobbs (2012) point to the inadequacy of using a linear urban-rural gradient as many studies viewed habitat availability and quality along 'city center – city edge – peri-urban' gradients, or 'urban – suburban – natural' area gradients, whereas modern cities exhibit more complex spatial patterns. Modern mega-cities and urban areas exhibit fractal-like patterns, characterized by multiple hierarchical clusters (Batty 2008) and expand non-linearly by presenting leapfrogging type of dynamics (Benguigui and Czamanski 2004). For a better understanding of the function of urban ecosystems Ramalho and Hobbs (2012) call on expanding beyond the traditional view of city structure and dynamics and propose to consider spatial patterns of the urban matrix and their temporal dynamics. Thus, the distance of a patch from a more urbanized city center may be less relevant than the distance of an open space from the urban edge. However, neighborhood relationships among open spaces, their size, connectivity and evolution through time may all be factors important for determining the properties of the urban-ecological framework.

In the light of the above mentioned discussion we studied species richness patterns in relation to the complex environment of two urban areas that differ in climate, topography, age and structure of the urban landscape. We tested if patch size and distance from the urban edge affects vascular plant species richness in urban open spaces. In particular we focused on the following hypotheses:

H1: The total number of vascular plant species in urban open spaces increases with increasing patch size.
H2: The number of native vascular plant species in urban open spaces increases with increasing patch size.
H3: The proportion of native vascular plant species in urban open spaces increases with increasing patch size.
H4: The total number of vascular plant species in urban open spaces decreases with increasing distance from the urban edge.
H5: The number of native vascular plant species in urban open spaces decreases with increasing distance from the urban edge.
H6: The proportion of native vascular plant species in urban open spaces decreases with increasing distance from the urban edge.

Hypotheses H1, H2, H4 and H5 explicitly assess whether the species respond to either distance from the urban edge or to patch size. Hypotheses H3 and H6, however, implicitly consider the response of the native species to the urban landscape. In spite of the different study areas we explore whether ecological principles can outweigh the impact of geographical position.

2 Methods

2.1 Study Areas

To test our hypotheses the cities of Hannover, Germany and Haifa, Israel were selected. Hannover is located in the north of Germany (city center: 52°23′ N, 9°42′ E) in humid Central Europe. Hannover represents an old city (founded in medieval times) along the banks of a low land river (Leine) in with a human population of more than half million (average of 2,576 inhabitants per km^2). The township of Hannover covers an administrative area of 204.1 km^2 with 37 % built-up, 16 % traffic, 15 % agriculture, 14 % open space, 11 % forest, 4 % water bodies and 4 % other land uses (Table 1) reaching altitudes from 44 m to 118 m a.s.l.. The city is located in the warm temperate climate zone with a mean annual temperature of 9.6 °C and mean precipitation of 661 mm (DWD 2013a, b).

Haifa is located in North-western Israel (32°48′ N, 35° E) on the Carmel Ridge, rising up to an elevation of 480 m above the Mediterranean Sea (Table 1). The jurisdictional area of the city is 63.6 km^2, and mean population density is 4,246 inhabitants per km^2. The climate is Mediterranean, and mean annual precipitation varies with elevation. At the lower elevations, at sea level, mean annual precipitation is close to 540 mm, and at the higher elevations it is approximately 685 mm (IMS 2011). The majority of the soils are formed on carbonate sedimentary lithology, dominated by limestone and chalk-marls.

2.2 Selection of Study Sites Within the Study Areas

A prerequisite to identification of the open spaces within the city was delineation of the urban area. We defined the urban edge as the border between the urbanized area with a high amount of sealed surface, and the open landscape with a much lower percentage of sealed soil. In Hannover we located the urban edge using detailed land cover classes of an digital landscape model (DLM) (LGN 2007). In Haifa the urban edge was delineated from aerial images obtained during 2008 and 2012. This city borders to the west and southwest large tracts of undeveloped maquis and planted pine forests, which mostly reside in the Carmel National Park. To the east the study area was truncated by an Industrial zone which beyond is followed by large intensively cultivated fallow crop fields. Within the urban edge all urban open spaces were identified by a low amount of sealed surface (generally < 25 %). Open spaces that had a mutual border were considered a single patch. The areas of each open space and its distance from the centroid of a polygon or, respectively, from the nearest patch border point was measured to the urban edge using GIS.

Study sites were chosen using a stratified random approach (see Underwood 1997) with three size classes (Hannover: >0.5–2 ha, >2–6 and >6–100 ha; Haifa: >0.1–1.6 ha, >1.6–37.5 ha, >37.5 ha) in combination with three distance classes

Table 1 Comparison of the study areas Hannover, Germany and Haifa, Israel

	Hannover	Haifa
Administrative area	204.1 km^2	63.6 km^2
Land use		
Built-up	37 %	37 %
Traffic	16 %	14 %
Agriculture	15 %	0 %
Open space	14 %	3 %
Forest	11 %	30 %
Water bodies	4 %	0 %
Others	4 %	16 %
Climate	Warm temperate	Mesic mediterranean
Mean annual temperature	9.6 °C	20.3 °C
Mean annual precipitation	661 mm	600 mm
Altitude	44–118 m a.s.l.	0–480 m a.s.l.
Soils	Mostly anthropogenic soils	Limestone accompanied by terra-rosas and chalk-marl accompanied by light-rendzinas
Inhabitants (2011)	525,875	270,348
Inhabitants/km^2	2,576	4,246

(Hannover: >50–1,000 m, >1,000–2,000 m and >2,000 m; Haifa: ≤300 m; >300–2,000 m, >2,000 m). From each category up to 5 open spaces were randomly chosen as study sites (Figs. 1 and 2). In Hannover 32 study sites were investigated. These contain 4 study sites for each size and distance class combination, but none for medium sized open spaces in the largest distance class. Habitat types found in these patches include forests, parks, cemeteries, allotments, ruderal sites and combinations of these. In Haifa 37 study sites were surveyed in which maquis, garrigue, ruderal site, park and grasslands were found.

2.3 Survey of Vascular Plant Species

All vascular plant species growing in each open space were recorded in Hannover from June to August 2011 and March to May 2012 and in Haifa during spring 2011 and spring 2013. Surveys were conducted afoot by a single person (SM for Hannover, DK for Haifa) and each open space was completely covered. Study sites were divided into dominant vegetation structures as they were surveyed (e.g. beech forest, lawn). Every vegetation structure was visited several times during the growing period to record all species. Ornamental species and self-established native and non-native species were recorded whereas submerse plant species were neglected. The nomenclature follows Buttler and Hand (2008) for wild growing

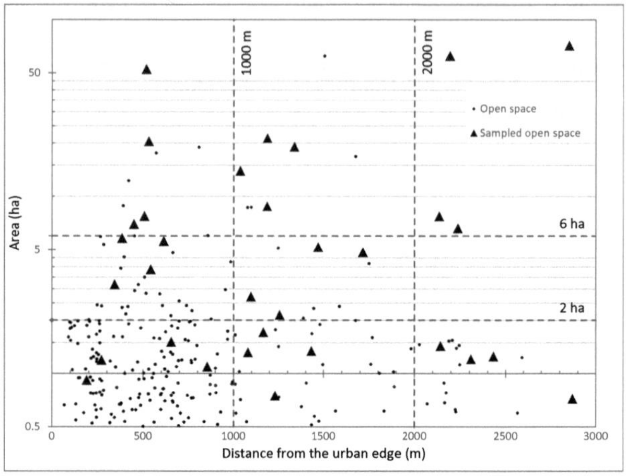

Fig. 1 Distribution of open spaces appropriate for selection within the studied area in Hannover. *Triangles* represent the sites chosen for sampling (note the log-scaled y-axis)

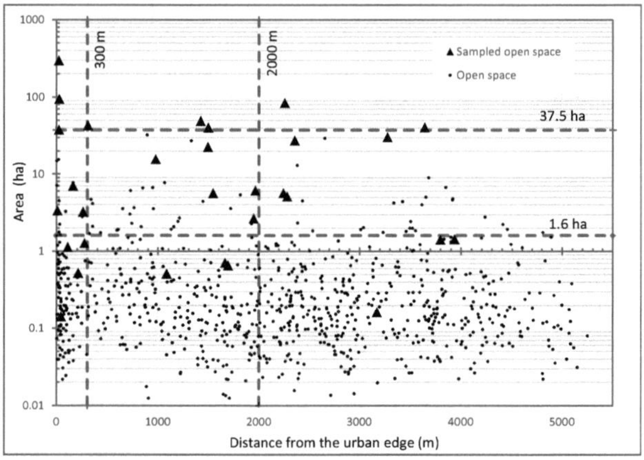

Fig. 2 Distribution of open spaces within the studied area in Haifa. *Triangles* represent the sites chosen for sampling (note the log-scaled y-axis)

species and Zander and Erhardt (2008) for ornamentals in Hannover. In Haifa the nomenclature follows Feinbrun-Dothan and Danin (1991).

2.4 Data Analysis

To test the six hypotheses total species number and the sub-groups into which they were classified – natives and proportion of natives – were designated as dependent variables. The decision whether a species is native or non-native to the region was based on the classification of Garve (2004) for Hannover and Feinbrun-Dothan and Danin (1991) for Haifa. The number of non-native species includes established and not established non-natives as well as ornamental species. The proportion of native species was calculated as the number of natives of all the species found in a patch.

Patch size and distance from the urban edge were the independent variables used and their values were calculated with a GIS. To linearize the relationship between the dependent and independent variables data were log-log transformed. Hypotheses H1-H3 were tested by regressing the dependent variables (number of species, number of native species and proportion of native species) against open space size. Hypotheses H4-H6 were tested by regressing the dependent variables against distance of open space from the urban edge.

3 Results

In Hannover 1,372 plant species were found in the open spaces surveyed, of which 577 (58 %) are natives and 795 (42 %) non-natives. In Haifa the total number of 363 species was found in urban open spaces, of which 334 (92 %) are considered natives and 29 (8 %) non-natives.

The patches of open space embedded in the urban matrix can be viewed as islands of suitable habitat, separated by a hostile background. If this is the case indeed, one can speculate that the relationship between patch size and the number of species will follow the commonly observed 'Species-Area curve' (Preston 1962). Accordingly, the data was log-log transformed, i.e., both the number of species and patch size were log-transformed, in an attempt to linearize the relationship between the number of species and patch size. Additionally, both within the framework of the Theory of Island Biogeography (MacArthur and Wilson 1967) and modeling of dispersal success (Levin et al. 2003), number of species or individuals is commonly viewed as decaying with distance from source. Consequently, we also log-transformed distance following the same rationale presented above.

The analyses revealed that the total number of species increased with increasing patch size and that this relationship was found to be significant both in Hannover and in Haifa (Table 2, Figs. 3 and 4). Similarly, the numbers of native species increase significantly with increasing patch size in Hannover and Haifa (Table 2, Figs. 5 and 6). In Haifa, however, the explanatory power of patch size, i.e., the percent of the variance associated with it, was lower compared to the Hannover case, as is evident by the lower adjusted R^2 values. Additionally, each change in a unit area in Hannover has a greater average effect on the expected number of species, as is evident by the higher b-coefficients obtained. Nevertheless, no significant influences were found between the proportion of natives and patch size (H3).

Additionally the effect of decreasing species numbers with increasing distance to the urban edge was not found to be significant (H4-H6). Results show no relation between the total number, the number of native or the proportion of native species to distance from the urban edge in both study areas.

Table 2 Linear regressions calculated for the dependent factors total species number, number of native species, and proportion of native species in combination with independent factors area and distance from the urban edge for Hannover and Haifa

		Intercept a			Slope b			
	Study area	Value	SE	P	Value	SE	P	$AdjR^2$
Log (total number of species) ~ log (area)	Hannover	2.23	0.03	<0.001	0.26	0.04	<0.001	0.51
	Haifa	1.02	0.14	<0.001	0.15	0.03	<0.001	0.41
Log (number of native species ~ log (area)	Hannover	2.07	0.02	<0.001	0.22	0.02	<0.001	0.72
	Haifa	0.8	0.19	<0.001	0.17	0.04	<0.001	0.33
Log (proportion native species) ~ log (area)	Hannover	1.84	0.02	<0.001	−0.04	0.03	n.s.	0.01
	Haifa	0.85	0.06	<0.001	0.02	0.01	n.s.	0.03
Log (total number of species) ~ log (distance from urban edge)	Hannover	2.56	0.36	<0.001	−0.05	0.12	n.s.	−0.02
	Haifa	1.80	0.20	<0.001	−0.02	0.06	n.s.	−0.02
Log (number of native species) ~ log (distance from urban edge)	Hannover	2.21	0.26	<0.001	0	0.08	n.s.	−0.03
	Haifa	1.89	0.26	<0.001	−0.08	0.08	n.s.	0
Log (proportion native species) ~ log (distance from urban edge)	Hannover	1.65	0.18	<0.001	0.05	0.06	n.s.	0
	Haifa	1.05	0.06	<0.001	−0.03	0.02	n.s.	0.05

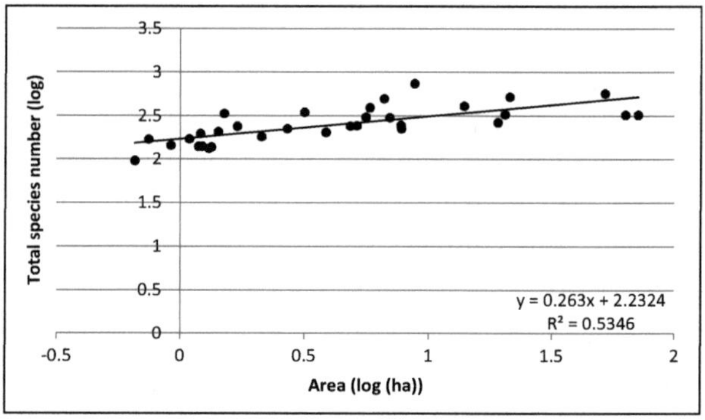

Fig. 3 Hannover – the relationship between the total species number and area (note the log-log transformation)

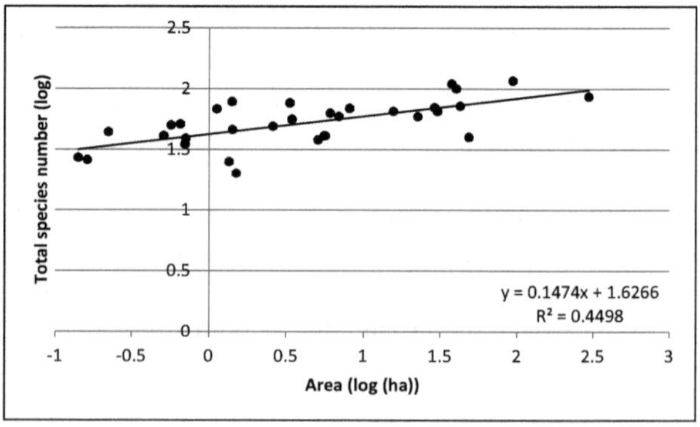

Fig. 4 Haifa – the relationship between the total species number and area (note the log-log transformation)

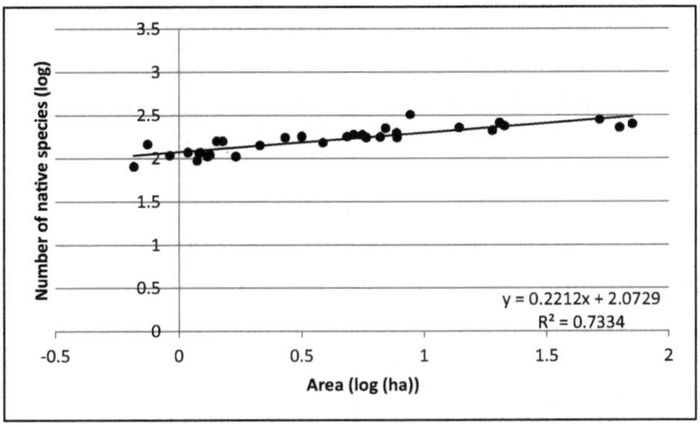

Fig. 5 Hannover – the relationship between the native species number and area (note the log-log transformation)

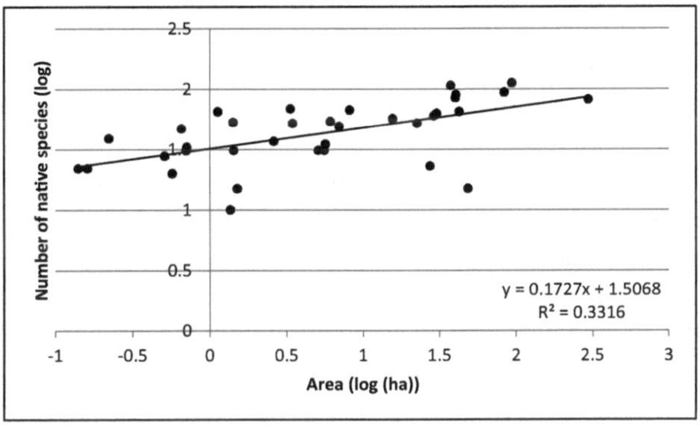

Fig. 6 Haifa – the relationship between the native species number and area (note the log-log transformation)

4 Discussion

As described above, Hannover and Haifa are dissimilar in many geographical properties. In addition, the land use and cultural history left their signature on the studied systems. The proportion of agricultural land within the boundaries of Hannover was close to 15 % whereas in Haifa there are no agricultural areas. When comparing the results of the analyses, however, the broad similarities of the dependent variable responses are striking, as in both urban landscapes the only explanatory variable found to be significant was patch size. This was demonstrated despite fundamental differences between Hannover and Haifa with respect to climate, topography and composition of the flora.

The urban matrix can be viewed within the framework of the Theory of Island Biogeography (MacArthur and Wilson 1967), and therefore it can be expected that the natural and semi-natural areas surrounding the cities provide the "mainland" species pool. Similarly, the open spaces within the city are the "islands" that are subjected to extinction and colonization processes. Therefore, it could be hypothesized that species numbers within open spaces are a function of patch size and distance from the urban edge. Our findings are in only a partial agreement with this theory as was indicated by the fact that distance from the urban edge was not found to be a significant factor. Results which identified the relationship between patch size and number of species were also obtained by others (Knapp et al. 2008; Meffert and Dziock 2012; Mörtberg 2001; Bolger et al. 1997), and these findings correspond with the concept of the species-area relationship (Preston 1962).

Additionally, the hypothesis that increasing patch size will result in increasing proportions of native vascular plant species did not yield any significant relationships. This may be due to the very different types of open spaces selected for investigation. The Hannover study sites included for example forests, parks, and cemeteries. Forests generally contain lower percentages of native species compared to higher values in parks or cemeteries, mostly due to ornamental species. The fact that our third response variable, the proportion of native species in open spaces, did not respond to either size or distance of a patch from the city's border indicates that the underlying assumption, that the overall pool consists of a pool of native species available from the natural areas surrounding the city, and a pool of non-native species from cultivated areas in the inner parts of the city, may be incorrect. Alternatively, the impact of the cultivated areas may not be strong enough to be detected.

In both cities distance from the urban edge was not a factor explaining species richness. This does not conform to many of the previous studies which investigated species richness patterns along urban to rural gradients. McKinney (2008) reviewed 17 studies related to plant species richness along an urban to rural gradient and 13 of the 17 studies report an association with the gradient, where 9 of the studies indicate peak diversity at the intermediate values of the gradient, one study reported peak values at the rural end of the gradient and in three studies peak values appear at the urban edge of the gradient. Other studies also demonstrated this relationship (Hope et al. 2003; Huste and Boulinier 2007; Bolger et al. 1997). The lack of relationship found in the current study may be explained by the fact that immigration processes may have a minor relevance in cities and that the urban matrix is more permeable to vascular plant dispersal than expected. This may be particularly true for cities which are heterogeneous and which are also characterized by many small patches that may serve as stepping stones for dispersal. For example Wania et al. (2006) studied flora richness in urban and agricultural landscapes of central Germany and concluded that the major driver for species richness is landscape heterogeneity, particularly of small patches associated with different land use types. In spite of this discussion the Theory of Island Biogeography does not work for the influence of distance from the urban edge on total and native vascular plant species numbers and the proportion of native species.

Therefore, we conclude from our preliminary results that city planning should focus on the creation and conservation of large open spaces, and also on providing a mosaic of different habitats within and between open spaces in order allow for a high urban biodiversity. The location of open spaces with respect to their distance from the urban edge seems to be of minor relevance, although the combined effect of patch size and the distance has not been investigated yet. Other factors, not investigated in this study may be of considerable influence on vascular plant species richness as well, for example the mosaic of different habitats within and between open spaces, disturbance characteristics and age of habitat.

Acknowledgements The presented investigations are part of the research project "Cities, nature and life between them: the dynamics of human and natural ecosystems and the interrelationships between them", which is conducted by the Institute of Environmental Planning, Leibniz Universität Hannover, Germany and the Technion – Institute of Technology –, Haifa and the University of Haifa, Israel. This joint research project was financially supported by the State of Lower Saxony, Hannover, Germany to SR, RP and DM, and was partially financed by research grant 666/10 from the Israeli Science Foundation to DM. Besides the presented investigations of species richness of vascular plants, the research project focuses on birds as well as mammals.

References

Batty, M. (2008). The size, scale, and shape of cities. *Science, 319*, 769–771.

Benguigui, L., & Czamanski, D. (2004). Simulation analysis of the fractality of cities. *Geographical Analysis, 36*, 69–84.

Blair, R. B. (1996). Land use and avian species diversity along an urban gradient. *Ecological Applications, 6*(2), 506–519.

Bolger, D. T., Scott, T. A., & Rotenberry, J. T. (1997). Breeding bird abundance in an urbanizing landscape in coastal southern California. *Conservation Biology, 11*(2), 406–421.

Buttler, K., & Hand, R. (2008). *Liste der Gefäßpflanzen Deutschlands. Kochia* (p. 107). Berlin: Gesellschaft zur Erforschung der Flora Deutschlands.

Drayton, B., & Primack, R. B. (1996). Plant species lost in an isolated conservation area in Metropolitan Boston from 1894 to 1993. *Conservation Biology, 10*(1), 30–39.

DWD. (2013a). Mittelwerte 30-jähriger Perioden. Mittelwerte für den Bezugsstandort am Ende der Referenzperiode 1981–2010. Niederschlag. http://www.dwd.de. Accessed 22 Mar 2013.

DWD. (2013b). Mittelwerte 30-jähriger Perioden. Mittelwerte für den Bezugsstandort am Ende der Referenzperiode 1981–2010. Temperatur. http://www.dwd.de. Accessed 22 Mar 2013.

Feinbrun-Dothan, N., & Danin, A. (1991). *Analytical flora of Eretz-Israel.* Jerusalem: Cana Publishing House.

Garve, E. (2004). *Rote Liste und Florenliste der Farn- und Blütenpflanzen in Niedersachsen und Bremen.* (Informationsdienst Naturschutz Niedersachsachsen 24 (1/04)). Hildesheim, Germany: Niedersächsisches Landesamt für Ökologie.

Guntenspergen, G., & Levenson, J. (1997). Understory plant species composition in remnant stands along an urban-to-rural land-use gradient. *Urban Ecosystems, 1*(3), 155–169.

Hope, D., Gries, C., Zhu, W., Fagan, W. F., Redman, C. L., Grimm, N. B., Nelson, A. L., Martin, C., & Kinzig, A. (2003). Socioeconomics drive urban plant diversity. *Proceedings of the National Academy of Science of the United States of America, 100*(15), 8788–8792.

Huste, A., & Boulinier, T. (2007). Determinants of local extinction and turnover rates in urban bird communities. *Ecological Applications, 17*(1), 168–180.

IMS. (2011). *Climatic Atlas of Israel for physical and environmental planning and design.* Israel Meteorological Service. (In Hebrew). http://www.ims.gov.il/IMS/CLIMATE/ClimaticAtlas. Accessed 7 Mar 2013.

Knapp, S., Kühn, I., Mosbrugger, V., & Klotz, S. (2008). Do protected areas in urban and rural landscapes differ in species diversity? *Biodiversity and Conservation, 17*(7), 1595–1612.

Kühn, I., Brandl, R., & Klotz, S. (2004). The flora of German cities is naturally species rich. *Evolutionary Ecology Research, 6*, 749–764.

Levin, S. A., Muller-Landau, H. C., Nathan, R., & Chave, J. (2003). The ecology and evolution of seed dispersal: A theoretical perspective. *Annual Review of Ecology, Evolution, and Systematics, 34*, 575–604.

LGN. (2007). ATKIS (Amtlich Topographisch-Kartographisches Informationssystem) Basis DLM (Digitales Landschaftsmodell) im Maßstab 1:10.000 bis 1:25.000.

MacArthur, R. H., & Wilson, E. O. (1967). *The theory of island biogeography* (Vol. 1). Princeton: Princeton University Press.

McKinney, M. L. (2008). Effects of urbanization on species richness: A review of plants and animals. *Urban Ecosystems, 11*, 161–176.

Meffert, P., & Dziock, F. (2012). What determines occurrence of threatened bird species on urban wastelands? *Biological Conservation, 153*, 87–96.

Melles, S., Glenn, S., & Martin, K. (2003). Urban bird diversity and landscape complexity: Species environment associations along a multiscale habitat gradient. *Conservation Ecology, 7*(1), 22.

Mörtberg, U. M. (2001). Resident bird species in urban forest remnants; landscape and habitat perspectives. *Landscape Ecology, 16*, 193–203.

Niemelä, J. (1999). Ecology and urban planning. *Biodiversity and Conservation, 8*, 119–131.

Preston, F. W. (1962). The canonical distribution of commonness and rarity: Part I. *Ecology, 43*, 185–215.

Ramalho, C. E., & Hobbs, R. J. (2012). Time for a change: Dynamic urban ecology. *Trends in Ecology & Evolution, 27*, 179–188.

Ricklefs, R. E., & Miller, G. L. (2000). *Ecology* (4th ed.). New York: W. H. Freeman and Company.

Snep, R. P. H., Opdam, P. F. M., Baveco, J. M., WallisDeVries, M. F., Timmermans, W., Kwak, R. G. M., & Kuypers, V. (2006). How peri-urban areas can strengthen animal populations within cities: A modeling approach. *Biological Conservation, 127*(3), 345–355.

Sukopp, H., Numata, M., & Huber, A. (1995). *Urban ecology as the basis of urban planning*. The Hague: SPB Academic Publishing.

Underwood, A. J. (1997). *Experiments in ecology: Their logical design and interpretation using analysis of variance*. Cambridge: Cambridge University Press.

Wania, A., Kühn, I., & Klotz, S. (2006). Plant richness patterns in agricultural and urban landscapes in Central Germany—spatial gradients of species richness. *Landscape and Urban Planning, 75*(1), 97–110.

Wickham, J., O'Neill, R., & Jones, K. (2000). Forest fragmentation as an economic indicator. *Landscape Ecology, 15*, 171–179.

Zander, R., & Erhardt, W. (2008). *Handwörterbuch der Pflanzennamen. Dictionary of plant names = Dictionnaire des noms de plants* (18th ed.). Stuttgart: Eugen Ulmer.

Future Suburban Development and the Environmental Implications of Lawns: A Case Study in New England, USA

Daniel Miller Runfola, Colin Polsky, Nick Giner, Robert Gilmore Pontius Jr., and Craig Nicolson

Abstract Lawns cover more land than irrigated corn in the United States according to the most recent estimates (Milesi et al. 2009). The associated ecological ramifications – such as habitat fragmentation, water quality and availability – may be far-reaching. The way lawns are maintained, especially intensive fertilization and watering, also presents risks for water use and quality, nutrient cycling, urban climate regimes, and even human health. However, the lack of broad-extent, high-resolution land cover data has limited the ability of researchers to measure or project the extent of lawns. In this chapter, we first produce a high resolution (0.5 m) land-cover classification to quantify existing lawn extent for the year 2005 in the Plum Island Ecosystem (PIE), a collection of 26 suburban towns northeast of Boston, MA, USA. We then use this dataset in conjunction with the GEOMOD land-change model to project lawn extent under two scenarios of urban growth for the year 2030. We find that in 2005, 76 km^2 of lawn – defined as grass on residential land – existed in the PIE study region. Under a Current Trends scenario, we project residential lawns may increase by 7.0 % to 81 km^2, while under a Smart Growth scenario we project a 1.6 % increase to 77 km^2. We estimate this could result in up to 61 million additional liters of annual water use under the Current Trends scenario, and 14 million under Smart Growth, putting additional stress on utilities that already face regular water shortages.

D.M. Runfola (✉)
National Center for Atmospheric Research, University of Colorado at Boulder, Boulder, USA
e-mail: dan@danrunfola.com

C. Polsky • N. Giner • R.G. Pontius Jr.
Clark University, Worcester, USA
e-mail: cpolsky@clarku.edu; nginer@clarku.edu; rpontius@clarku.edu

C. Nicolson
University of Massachusetts Amherst, Amherst, USA
e-mail: craign@eco.umass.edu

D. Malkinson et al. (eds.), *Modeling of Land-Use and Ecological Dynamics*, Cities and Nature, DOI 10.1007/978-3-642-40199-2_7,

1 Introduction

1.1 Research Objective

It is estimated that between 45,000 and 96,000 km^2 of lawns are irrigated each year in the United States, making it the single largest irrigated crop in the country (the second largest, corn, occupies approximately 39,000 km^2) (Milesi et al. 2009). As the extent of urban and residential land use continues to expand the total coverage of lawn is also expected to increase. While ample research has been conducted on the impacts of broad-scale agricultural land change on undeveloped environments, little literature exists on the impacts of broad-scale residential land change. However, similar to agricultural fields the presence and maintenance of lawns have a number of environmental ramifications including the modification of water use regimes, habitat fragmentation, water quality, soil profiles, runoff, water biochemistry, and human health (Nielson and Smith 2005; Robbins and Birkenholtz 2003).

Assessing and projecting the extent of lawns is thus valuable for urban planning efforts as cities consider different growth strategies (House-Peters and Chang 2011; Runfola et al. 2013). The nature of this planning, typically concerning neighborhood or even parcel-level dynamics, requires maps of present and future lawn cover at fine spatial resolutions (Robbins and Birkenholtz 2003). However, data availability has limited the ability of researchers to analyze or project the quantity of lawn at fine spatial scales, resulting in estimates derived from coarse-resolution data (e.g., 1 km^2) (Milesi et al. 2005). In response to the growing need for fine spatial resolution estimates – and projections – of lawn land cover, this paper examines the research question: "*What is the current extent of lawn in suburban Boston, and how much additional lawn might be expected under different scenarios of (sub)urban growth?*"

1.2 The Potential Impacts of Large-Scale, Resource-Intensive Lawn Care

In the United States over the past several decades, the physical pattern of urban growth has favored geographical expansion rather than increasing density, resulting in what is often referred to as "suburban sprawl," with an emerging pattern of "exurbs" and "boomburbs" (Lang et al. 2005). Perhaps the most common biophysical feature of these suburban landscapes is lawn (Hanlon et al. 2010). With increases in lawn area likely come increases in the inputs (chemicals, water, energy, dollars, time) required to maintain these lawns (Robbins 2007; Robbins and Birkenholtz 2003; Robbins and Sharp 2003). These inputs, and their associated environmental ramifications, have led an increasing number of researchers to examine the lawn landscape as a growing human-environment concern (Giner et al. 2013; Harris et al. 2012a, 2012b; Larson et al. 2009; Robbins 2007; Roy Chowdhury et al. 2011; Runfola et al. 2013; Zhou et al. 2008).

The broad-scale ecological concerns associated with lawn care remain largely a matter of speculation and conventional wisdom as most research regarding lawns has been conducted at small spatial scales. Moreover, there have been few if any studies projecting lawn cover patterns into the future, a critical component for urban planners seeking to incorporate a more nuanced understanding of lawn care into design efforts. To tackle these challenges and understand how lawn care may influence a watershed in the future will require not only data covering a broad extent and at a fine resolution, but also means for projecting patterns of (sub)urban growth to future time period(s) (Milesi et al. 2005; Robbins and Birkenholtz 2003).

Even though we are unaware of any published studies projecting future lawn extent and associated environmental consequences, researchers have examined various dimensions of historical effects of lawn care including water use, nitrogen processing, carbon removal, wildlife fragmentation, atmospheric conditions, water quality, and health. The relationship between water use and lawn maintenance has been a focus of much research (Domene and Saurí 2006; Domene et al. 2005; Mayer and DeOreo 1999; Milesi et al. 2005). In 2006, Domene and Saurí used used a sample of 532 households in Barcelona to examine water use. They found that while the link between the size of a yard and water use was not significant, garden design which necessitated high water use was significantly ($\alpha = 0.1$) and positively correlated with water use during all seasons and across both high and low density residential units. Using Landsat TM imagery (30 m spatial resolution), Wentz and Gober (2007) found that xeric landscaping in Phoenix, AZ had a relatively small impact on water use, contrary to conventional wisdom(see Larson et al. 2009). However, Wentz and Gober also identified a significant positive relationship between mesic landscaping and residential water use (a 1 % increase in mesic vegetation increased annual water use by ~4,000 l [~1,000 gal]). Also in Phoenix, Balling (2008) found that census block groups with a higher proportion of mesic landscape had more sensitivity to atmospheric conditions.

The quantity of water use is not the only environmental challenge brought on by lawn management. Water quality may also be affected through nitrogen and phosphorus runoff from fertilizer application, which poses risks to the health of humans, plants and animals (Nielson and Smith 2005). It is estimated that, circa 1999, 62 million kg (136 million pounds) of pesticides are applied to urban lawns and gardens annually in the United States (Nielson and Smith 2005). Further, over 50 % of nitrogen inputs into some (sub)urban watersheds have been attributed to lawn fertilization (Groffman et al. 2004). With more fertilizer being used for lawns per square meter than for food production (Carrico et al. 2012; Robbins and Sharp 2003), it is no surprise that reducing over-fertilization has been identified as a strategy for mitigating the impact of lawns (Carrico et al. 2012; Fissore et al. 2011).

Impacts of intensive lawn maintenance on biogeochemical cycling may also be large enough to merit inclusion in understanding local or regional atmospheric dynamics (Karl et al. 2001; Milesi et al. 2005). First, the annual vegetative growth associated with US lawns may be responsible for up to 17 Tg/y of carbon removal from the atmosphere (Milesi et al. 2005). Second, the mechanized nature of lawn care can lead to significant emissions of volatile organic compounds and other pollutants (Karl et al. 2001; Priest et al. 2000). Third, further evidence of the atmospheric impact of lawns comes from research examining the influence of urban vegetation – including evaporative cooling associated with lawn irrigation – on the urban heat island effect, illustrating the challenges of balancing water budgets with temperature amelioration (Chow et al. 2011; Gober et al. 2010).

In sum, resource-intensive lawn care presents a host of potential environmental and human health challenges. Thus it stands to reason that if such lawn care behaviors are widespread, then these rapidly growing suburban ecosystems present a problem of national significance. Yet, prior literature making this case is largely grounded in studies using either broad-extent and coarse-resolution, or small-extent and fine-resolution, data. Studies using broad-extent and fine-resolution data are needed. The project reported in this chapter produces and analyzes such a dataset in a (sub)urban community of Boston, MA, first to quantify the current extent of lawns, and second to project future lawn cover patterns under two scenarios of urban growth.

2 Methods and Data

2.1 Study Area

Developed land in Massachusetts increased approximately 59 % between 1972 and 1996 and suburbs in the city of Boston are some of the fastest growing in the United States (Krass 2003). From 2000 to 2009, Boston experienced population growth of nearly 10 % (US Census Bureau 2011). Population in metropolitan Boston has decreased over the past three decades, as its suburbs have experienced substantial development and increases in population (Tu et al. 2007; US Census 2013). The average number of acres per housing unit doubled in the Boston region between 1970 and 2000, and the median distance of new homes from central Boston grew concurrently (MAPC 2010).

This study focuses on an 1,150 km^2 area to the north of the city of Boston (Fig. 1): the Plum Island Ecosystems (PIE). The PIE study area is a National Science Foundation Long Term Ecological Research (LTER) site, comprising of 26 suburbanizing towns that intersect the Ipswich and Parker River Watersheds (http://ecosystems.mbl.edu/pie/). Two portions of the study area, the North Shore and North Suburban sub-regions, have experienced a quadrupling in their populations between 1970 and 2000 (MAPC 2010).

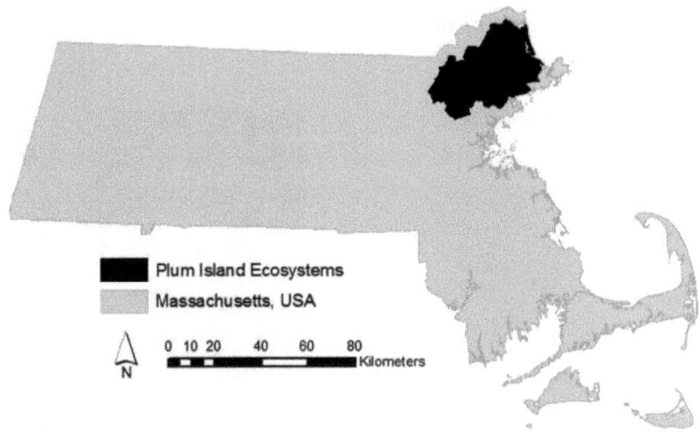

Fig. 1 The Plum Island Ecosystem (PIE) study area

2.2 *Data*

To produce a high quality, fine-resolution dataset of lawn cover, we rely upon remotely sensed data. MASSGIS (2011) provides digital 8-bit, four-band (blue, green, red and near-infrared) orthoimagery captured between April 9 and April 17, 2005 (leaf-off) with a 45 cm spatial resolution (MassGIS 2011). These data are received geometrically corrected and resampled to a 50 cm resolution using bilinear interpolation (Fig. 2). Digital ancillary data (2003 assessor's parcels, a wetlands layer, 1971 and 2005 land use, slope, and town boundaries) are acquired from MassGIS (2011). Census and roads data from the year 2000 are acquired from the US Census Bureau (www.census.gov). Information on both the Smart Growth and Current Trends projections were provided by the Metropolitan Area Planning Council (MAPC), a public policy group representing the Boston metro area.

2.3 *Quantifying Lawn Land Cover in 2005*

The first step of this research is to map existing lawn cover. To this end, object-based image analysis (OBIA) is used to produce a 0.5 m classification of lawn land cover across the entire PIE region. OBIA is a relatively new approach to image classification (Blaschke 2010; Blaschke et al. 2000). First, orthoimagery is segmented into "image objects" using a segmentation algorithm, in this case a multi-resolution image segmentation algorithm (Burnett and Blaschke 2003; Benz et al 2004). This segmentation algorithm groups contiguous pixels with similar characteristics (frequently spectral characteristics) into objects (Blaschke and Strobl 2001). Numerous new attributes of these objects, including mean normalized difference vegetation index (NDVI), mean albedo (the degree to which an object reflects light in all three bands), size, shape, and distance to adjacent objects are calculated to assist in the delineation of seven land cover classes: bare soil, coniferous, deciduous, turf grass, impervious, water, wetlands. In order to assign

Fig. 2 An example of the raw 0.5 m orthoimagery, and the 120 m resolution cells it is aggregated to

one of these classes to each object, a three step approach is taken. First, objects which fall into classes for which ancillary data is available (impervious, water, wetlands) are initially soft-classified based on what ancillary layer they fell into. Second, NDVI and brightness thresholds are qualitatively assessed for every class. These thresholds are used as conditions whereby objects are assigned to a particular class. During this step, objects classified on the basis of ancillary data are re-assessed, and those that fell outside of the thresholds for the class to which they were assigned have their classification removed. This process is performed hundreds of times for small subsets of the study area.

Remaining areas are classified by selecting representative objects as training sites to define all seven land cover classes and then classifying image objects based on these samples using a nearest neighbor algorithm. The classifications are then visually inspected for positional and classification accuracy, and manually edited to correct misclassified objects. An example output can be seen in Fig. 3.

Accuracy assessment is performed by hand digitizing thousands of polygons grown from points randomly dropped within each stratum. These hand digitized results are compared to the object based mapping product using a crosstabulation matrix. The proportion of areas which are the same in both the hand digitization and the map is calculated then weighted by the area of each class to produce an overall measurement of accuracy. More details are available in Polsky et al. (2012).

2.4 Model Comparison and Selection

There are a host of models and methods to project (sub)urban lawn growth as a result of residential growth. Even though no literature has utilized wall-to-wall, high resolution data to project future lawn extents, urban growth modeling has a rich tradition in the land change modeling community (Agarwal et al. 2002; Silva and Clarke 2002; Tobler 1970; Yang and Lo 2003). Numerous approaches

Fig. 3 An example of the 0.5 m mapping product output

to urban growth modeling have been produced, each presenting a number of tradeoffs (Agarwal et al. 2002). To assist in the selection of an appropriate model, Table 1 provides a summary of 13 spatially explicit land change models, summarized based on five dimensions:

1. Is the model deterministic or stochastic, i.e. does the model have a random component that can result in different outputs with the same inputs?
2. Is the model is dynamic or static, i.e., does the model explicitly allow the outcome at one point in time to influence the outcome at subsequent points in time?
3. Is the model mechanistic, i.e. does the model explicitly include interactions between individual parts of the system?
4. Is the model empirical, i.e., does the model utilize historic data to project future trends?
5. Is the model top-down or bottom-up, i.e., does the model project the total quantity of new residential land first and then spatially allocate it (top-down), or does it calculate the total quantity based on the spatial projection?

Each model also has a variety of key assumptions and characteristics. Rather than provide an exhaustive list, we have generalized these into five major categories:

1. *Past rates of change can be directly used to predict future patterns*. This is particularly important for empirical models which utilize historic information to project future scenarios (Runfola and Pontius Jr 2013). In many cases, this assumption is violated – i.e., if a country underwent a major political change or war.
2. *Spatial relationships between past and future growth are relevant.* Some models utilize spatial information to calibrate future projections – for example, in this paper we assume neighbors will have similar lawns.

Table 1 A comparison of published and available land change modeling packages. Costs are based on non-discounted, single seat licenses

			Model characteristics					
Model	Author (Year)	Assumptions/ characteristics	Deterministic and/or stochastic?	Dynamic and/or static?	Mechanistic and/or empirical?	Top-down and/or bottom-up?	Cost (USD)	Other information
CLUE-CR	Veldkamp and Fresco (1996)	1, 2, 3	Stochastic	Dynamic	Hybrid	Hybrid	0	A manual and all versions of the CLUE model can be downloaded from: http://www.ivm.vu.nl/
California Urban Futures (Second Generation) (CUF)	Landis and Zhang (1998)	1, 2, 3, 4	Stochastic	Dynamic	Hybrid	Top-down	No data	Requires ESRI Products
Land-Use Change Analysis System (LUCAS)	Berry et al. (1994)	1, 2	Stochastic	Both – can be set by user	Empirical	Bottom-up	0	Free, but requires UNIX and GRASS Experience: http://web.eecs.utk.edu/~lucas/
GEOMOD	Pontius et al. (2001)	1[a]	Deterministic	Both – can be set by user	Empirical	Top-down	1,250	Part of the IDRISI GIS Package http://www.clarklabs.org
SLEUTH	Silva and Clarke (2002)	1, 2, 3	Stochastic	Dynamic	Empirical	Bottom-up	0	http://www.ncgia.ucsb.edu/projects/gig/
Land Use Scanner (LUS)	Hilferink and Rietveld (1998)	1, 2	Stochastic	Dynamic	Hybrid	Top-down	0	Integrated into the Lumos Toolbox: http://www.lumos.info/
Environment Explorer (EE)	Engelen et al. (2003)	2	Both – can be set by user.	Dynamic	Mechanistic	Hybrid	0	More information can be found at: http://www.lumos.info/environmentexplorer.php

SAMBA	Castella et al. (2005)	5	Stochastic	Dynamic	Mechanistic	Bottom-up	No data	Calibration is very intensive
Land Transformation Model (LTM)	Pijanowski et al. (1997)	2, 3	Stochastic	Dynamic	Mechanistic	Top-down	0	Available at: http://ltm.agriculture.purdue.edu/default_ltm.htm
Behavioral Landscape Model (BLM)	Walker et al. (2004)	4, 5	Stochastic	Dynamic	Mechanistic	Bottom-up	No data	Calibration is very intensive
UPLAN	Johnston et al. (2003)	1, 2	Stochastic	Both – can be set by user	Mechanistic	Top-down	0	Requires ESRI Products http://ice.ucdavis.edu/project/uplan
UrbanSim	Waddell (2002)	2, 5	Stochastic	Dynamic	Mechanistic	Hybrid	0	Available as a standalone download: http://www.urbansim.org/
What if?	Klosterman (1999)	2	Deterministic	Both – can be set by user	Mechanistic	Bottom-up	2,495	Available at: http://www.whatifinc.biz

1 – Past rates of change can be used to predict future patterns ([a]Only if quantity input is based on empirical/historical trends)
2 – Spatial relationships between past and future growth are relevant
3 – Requires specific ancillary data
4 – Individual rationality and independence in decision making
5 – Actors chosen for agent calibration are representative of the population being studied

3. *Specific ancillary data is assumed to be relevant.* Rarely, models will assume a-priori the most important types of data for your analysis. It is important to consider if all data inputs really are relevant.
4. *Individual rationality and independence in decision making.* Many economic and agent-based models assume individual rationality and independence, which are not always true – are your actors behaviors "irrational" when compared to what you can model?
5. *Actors chosen for calibration are representative of the population.* Just as empirical land change models must make sure that past land change trends are helpful in informing future trends, so must models with mechanistic or agent-based approaches make sure that their calibration information is representative of their target population. For example, only interviewing policymakers and using them to represent all actors in a model would represent a critical assumption.

For our purposes, we prefer a model that produces replicable results, and require a model with a top-down approach. This is driven by the end-goal of this research – results which can be consistently communicated to policymakers. Replicable results allow both researchers and policymakers to compare outcomes more readily, while the top-down approach is required by this project as quantity projections of residential growth have already been provided. While many of the models listed here could potentially be modified to serve our needs, we chose GEOMOD as it both satisfies these criteria and is relatively simple to communicate to audiences with limited modeling experience.

2.5 Projecting 2030 Residential Development: GEOMOD

The second step of the analysis was to project the potential location of new residences under each scenario (Current Trends and Smart Growth). Ultimately, 120 m cells were chosen as the resolution at which to project new residential growth – smaller spatial scales were both technically limiting (there are approximately 4.5 billion 0.5 m pixels in the PIE region) as well as theoretically less interesting (projecting a single 0.5 m pixel of new urban growth is roughly akin to projecting a bird bath being added to a back yard) (Fig. 2). The results of aggregating lawn to the 120 m cell resolution can be seen in Fig. 4.

The quantity of new residential growth provided by the Boston MAPC is input into GEOMOD for each residential development scenario, which then spatially allocates the specified quantity according to three decision rules. The first decision rule determines the search radius around existing residential land cover to allocate new residential development. If this rule is used, it constrains growth spatially so that residential growth is near existing residential pixels. In the Current Trends scenario, development is allowed to occur anywhere on the landscape. In the Smart

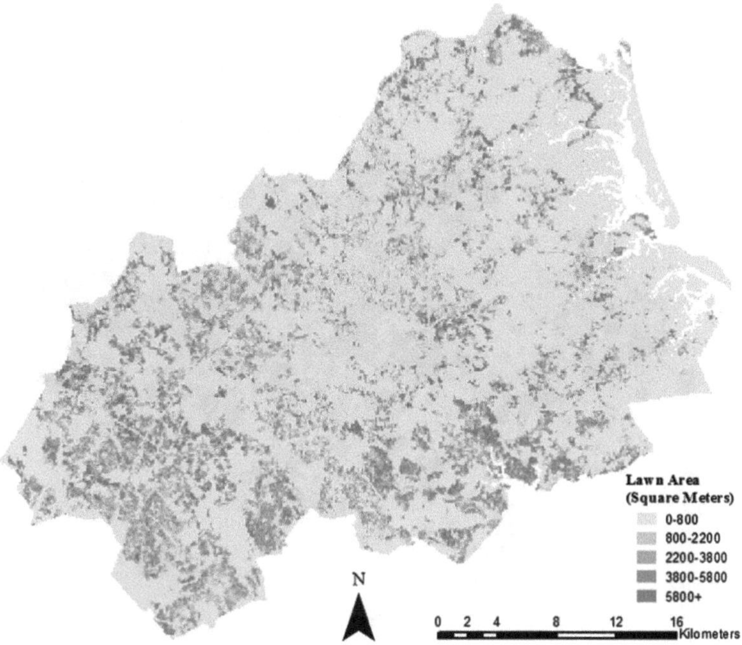

Fig. 4 The amount of lawn in each 120 m cell across the PIE region

Growth scenario, this decision rule is used to limit new growth to occur only in cells bordering existing residential areas.

The second decision rule is a binary restraint on development whereby some areas can be excluded from eligibility. In this study we excluded areas such as wetlands that are protected by the town and state and land use types that are deemed by us as ineligible for development such as power plants (Fig. 5). Remaining areas were deemed eligible.

The third decision rule determines how change is allocated within eligible locations for development. This is done on the basis of a suitability surface. We follow numerous authors in utilizing the distance to roads, past land use/cover, slope, and information on contemporary land use/cover to calibrate this suitability surface (e.g., Silva and Clarke 2002; Lo and Yang 2002; Yang and Lo 2003; Pinto et al. 2009). For both scenarios, GEOMOD crosstabulates each calibration surface with a map of 2005 residential land use – i.e., 1971 land use is crosstabulated to determine the land uses in 1971 that experienced the most residential growth between 1971 and 2005. A suitability surface is then created by computing the percent of each land use class in 1971 that transitioned to residential by 2005, and applying those percents spatially according to where each land use existed in 1971. This is performed for each calibration surface, and the suitability scores are then averaged to calculate a single suitability for every eligible 120 m pixel (Fig. 6). GEOMOD then searches within areas constrained by decision rules 1 and 2, and

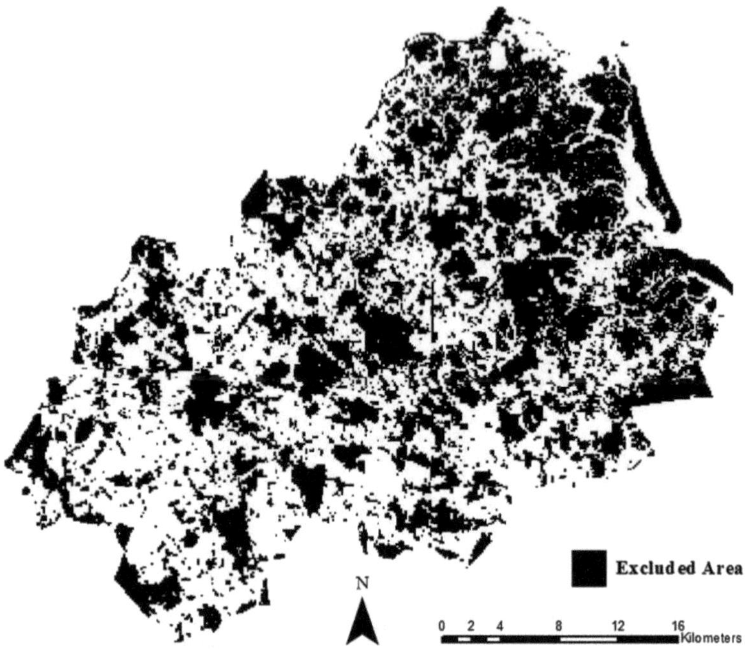

Fig. 5 Areas excluded from becoming new development in 2030

allocates new residential pixels on areas with the highest suitability until the specified area of new residential growth has been allocated.

2.6 Projecting Future Lawn Extent

After new residential change is allocated across the map, we project lawn cover. This step uses ordinary kriging, which is a technique that extrapolates based on spatial dependence. For each 120-m cell that is projected to become new residential development, we project the amount of that cell that will be lawn based on the amount of lawn in neighboring cells. This is done using a weighted linear combination (WLC) of each neighboring cell, in which cells farther away will tend to receive less weight. These weights are calibrated by examining how similar 2005 residences that fall a certain number of meters apart from one another are for a large number of distances until a variogram can be estimated. For estimations, omnidirectional variogram values for 12 lag distances (h) of 120 m are derived following Eq. 1 (Isaaks and Srivastava 1989):

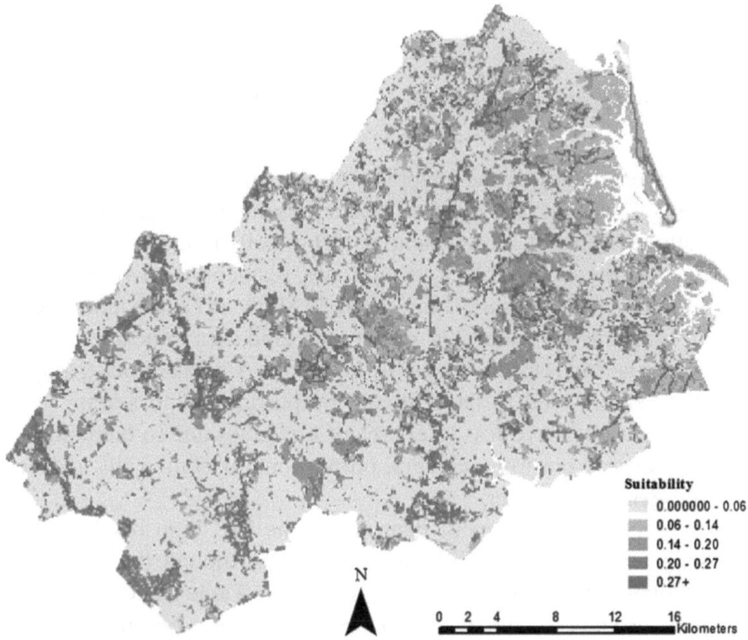

Fig. 6 The suitability surface generated based on historic growth patterns

$$2\gamma_h = \frac{1}{n_h} \sum_{P_{hij} \in N} \left(V_{hi} - V_{hj}\right)^2 \tag{1}$$

Where n_h is the number of paired 120 m cells at lag distance h, $\boldsymbol{N}$ is the set of all such pairs P_{hij}, V_{hi} is the value at paired value i, and V_{hj} is the value at paired value j. After Eq. 1 is calculated for all lag distances, a variogram is manually calibrated using a spherical model. Finally, weights are derived by solving:

$$\mathbf{W} = \mathbf{V}^{-1} * \mathbf{D} \tag{2}$$

where **W** represents a one dimensional (n + 1) by 1 weights matrix, $\mathbf{V}^{-1}$ is the inverted (n + 1) by (n + 1) matrix of all sampled variogram values derived on the basis of the estimated variogram, and **D** is a one dimensional (n + 1) by 1 matrix of distance values from each new 2030 residential cell unit to each 2005 residential cell derived from the estimated variogram.

3 Results

3.1 Mapping Historical Landcover in PIE

An example of results from the mapping effort, and example 120-m cells, are presented in Figs. 3 and 4. Each 0.5 m cell represented in Fig. 3 is aggregated to the 120 m cells seen in Fig. 4. Residential 120 m cells range from 0 % to 89.5 % lawn coverage. At the census block group scale, residential lawn coverage ranges from a minimum of 1.06 % to a maximum of 37.89 % turf grass land cover. At the town scale, it ranges from under 2 % of the total land area to nearly 12 % (Fig. 7). The total residential lawn in the region is 75.77 km^2. Census block groups located closer to town centers tend to have greater proportions of turf grass land cover, although they generally have less total area of turf grass due to their smaller sizes. As one moves away from town centers, proportions of coniferous and deciduous land cover are generally higher.

The 0.5 m accuracy assessment determined the PIE high resolution mapping product has an overall weighted accuracy of 80 %, though the accuracy of our reclassified "lawns" and "other" map is 94.11 % (see Table 2). Most of the error in the 7-category product is the result of confusion between the deciduous and wetlands map categories. It is believed this error is due to the mismatch between the aerial imagery methodology used to validate our maps and the survey work performed to create the wetlands map. The biggest source of confusion with the lawn category is bare soil. This confusion is largely due to the difficulty in spectrally distinguishing between sparsely vegetated areas, lawns, dead or dying grass, and bare soils.

3.2 Projecting New Residential Developments During 2005–2030 Using GEOMOD

Using the GEOMOD land change model, 73.1 km^2 of new residential development are distributed across PIE in the Current Trends scenario (Fig. 8). After accounting for legal and other restrictions on development, this residential development primarily took place in areas that had a gentle slope (<8 %), within 120 m of an existing road, and on forested areas. Over 20 % of the development occurs in three of the 26 towns, Ipswich, Billerica, and North Andover, three of the largest and relatively sparsely developed towns in the study area. By approximating the similarities of neighbors in terms of lawns in 2005 and applying this relationship to the new residential developments, in the Current Trends scenario it is estimated 5.3 km^2 of this new development, i.e. 7 % of the newly developed land would be residential lawn.

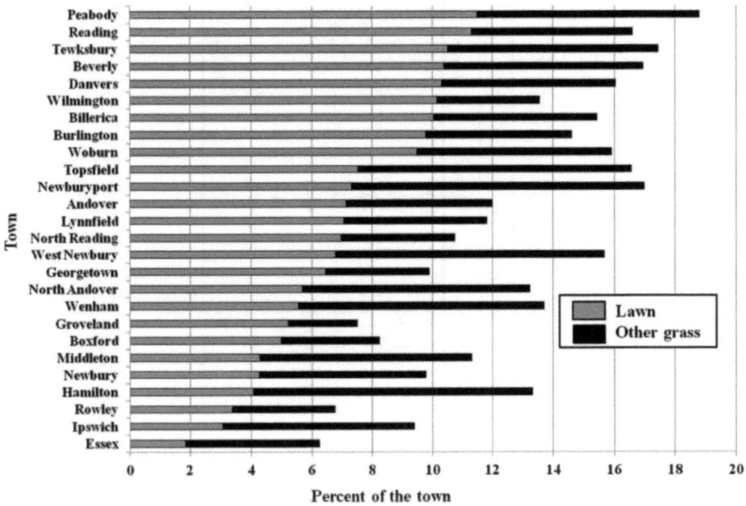

Fig. 7 The amount of lawn in each town across the PIE region, adopted from Giner et al. 2013

Under the Smart Growth scenario, 16 km^2 of new residential development are distributed (Fig. 9). While suitability was calculated in the same manner as in Current Trends, residential development was limited to areas near existing residential development. It was estimated that 1.2 km^2 of this new development, i.e. 7 % of the newly developed land would be residential lawn.

The total quantity of observed and projected lawn is summarized in Fig. 10. As prescribed by our model parameters, under the Smart Growth scenario urban growth all occurred next to existing residential areas, resulting in a much more clustered pattern of urbanization than in the Current Trends scenario. In both scenarios, very little development is projected to occur in the eastern part of PIE, due to a high distance to roads and a domination of the wetlands land cover category, a legally protected land use. The majority of development is projected to occur in the northern half of the study area, as the southern half is already highly developed (leaving little space for infill or new construction).

4 Discussion

4.1 Land Change Model

In addition to its deterministic and top-down nature, GEOMOD was desirable for this analysis for a number of reasons: it is fully controllable, transparent, simple to describe, and is explicit in its assumptions. However, many minor and major

Table 2 The population-adjusted proportional crosstabulation for the accuracy assessment of the Plum Island Ecosystems 0.5 m mapping product

	Photo interpreter								
		Bare Soil	Coniferous	Deciduous	Turf grass	Impervious	Water	Wetlands	Map total
Map	**Bare soil**	2.32	0.01	0.34	0.65	0.33	0.00	0.09	3.74
	Coniferous	0.04	9.77	2.45	0.37	0.07	0.06	0.07	12.83
	Deciduous	0.55	1.59	29.62	1.73	0.27	0.28	0.63	34.68
	Turf grass	1.29	0.23	0.88	9.87	0.41	0.04	0.02	12.73
	Impervious	0.49	0.06	0.30	0.17	13.61	0.02	0.00	14.64
	Water	0.02	0.02	0.09	0.01	0.00	2.65	0.18	2.97
	Wetlands	0.45	0.62	4.67	0.08	0.04	0.74	11.79	18.40
	Photo total	5.16	12.30	38.35	12.88	14.73	3.79	12.78	100.00

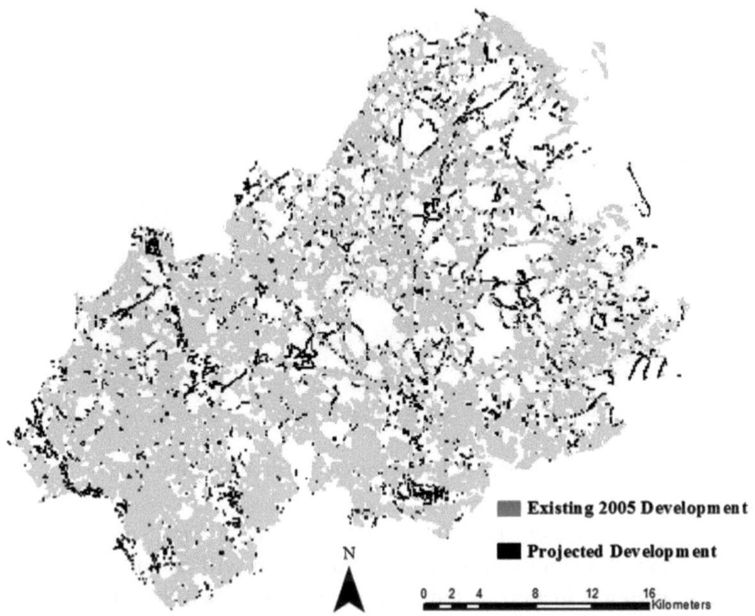

Fig. 8 Projected new residential development under the "Current Trends" growth scenario

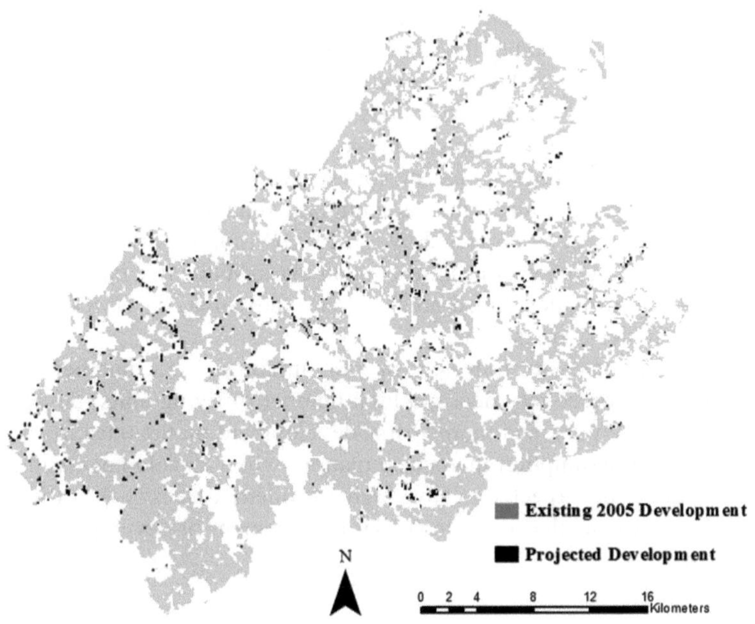

Fig. 9 Projected new residential development under the "Smart Growth" growth scenario

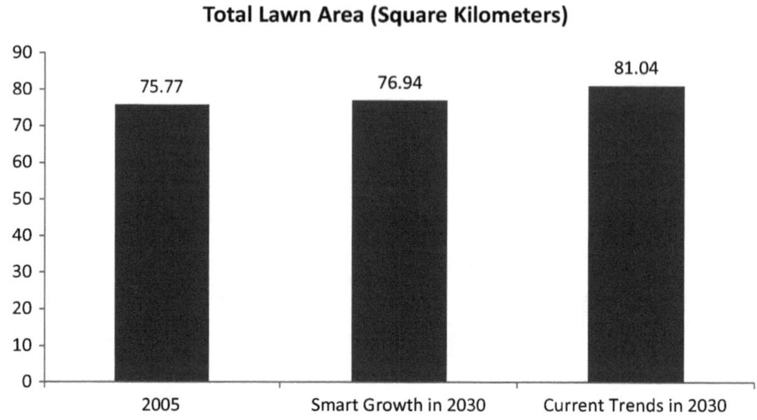

Fig. 10 A comparison of the amount of lawn projected in each scenario

assumptions of GEOMOD are reflected in this study. First, any area that was ineligible for new development at the first time step will be ineligible at all time steps – i.e., if an area is in a protected wetland in 2005, it will never become eligible for development. This prevents some changes that might actually be possible on the landscape – i.e., a protected wetland today may not be protected if legal frameworks change in the future.

Second, our particular implementation of GEOMOD assumes that spatial relationships matter. While not a part of the stock model, in our case we use spatial relationships to project the area of lawns in new residential developments. While spatial autocorrelation between 120 m pixels does exist for these attributes in the present (as measured by the variogram), there is no way to know if this assumption will extend into the future. In fact, some of our recent work suggests that more heterogeneity exists in these (sub)urban environments than is commonly thought (Harris et al 2012a).

Third, only gains can be projected by GEOMOD (not losses) – i.e., if a pixel is classified as urban, it can never become another class. While it is rare for developed areas to transition to undeveloped areas in our study area, GEOMOD would not account for such "loss" transitions.

These assumptions are reflected in our findings in a number of ways. For example, we project very dense (a high number of living units) developments very close to city centers, and more sparse development on the outskirts of towns. This is due to our assumption that space matters – new developments in existing highly urban areas are expected to have higher residential density than those on the outskirts of the city, as that was the spatial arrangement of development in 2005, our base year.

Table 3 The total amount of lawn, and associated water use, for 2005 and each scenario

	2005	Smart growth	Current trends
Square kilometers of lawn	75.7663	76.94 (+1.17) + 1.55 %	81.04 (+5.28) + 6.96 %
Estimated liters of water used for lawn maintenance	878,889,080	892,511,071 + 13,621,991	940,098,799 + 61,209,719

4.2 *Implications*

Each square meter of new lawn, and associated lawn management, leads to a wide array of potential environmental impacts ranging from water quality issues to wildlife fragmentation. For purposes of brief illustration, we link the projections of future lawn patterns with one environmental outcome variable: residential water consumption. Table 3 presents a comparison of our projected lawn water use for 2005 as well as both 2030 scenarios.

In terms of water use, using our estimates calibrated using one town in our study area (Runfola et al. 2013), we estimate that 879 million liters of water were consumed in PIE during 2005 for lawn irrigation. This is just 100 l less than the total water used for residential purposes by one of the towns in our study area, Ipswich, in 2005. In the Smart Growth scenario, this number increases to 893 million (+1.6 %), while under current trends this number increases to 940 million liters (+6.9 %). Such increases in water use may present concerns for two reasons. First, many towns in the region are already implementing demand-reduction measures, including price changes, new metering approaches, water-use restrictions, and incentives for homeowners to adopt high-efficiency technologies such as low-flow showerheads or rain barrels (Hill and Polsky 2007; Krahe et al. 2012; Glennon 2002, 2009; Ipswich Utilities 2011). As such, there may be limited additional achievable reductions in per capita demand. Second, this increase only incorporates one of many drivers of increased water use – that is, lawns. As such, the realized increase in aggregate water use in 2030 will surely be greater than what is projected here, even if modest additional reductions in per capita demand are achieved. Further, we do not account for any changes that might occur in the agricultural, commercial, or industrial sectors of the economy, only focusing on residential lawns and water use.

5 Conclusions

This chapter projects two scenarios of future (sub)urban growth in the Plum Island Ecosystem (PIE) Long Term Ecological Research site, 26 towns to the north of Boston, Massachusetts, USA. First, a high resolution (0.5 m) dataset is created for the entire study area, allowing the quantification of the extent of contemporary residential lawns. Second, the land change model GEOMOD is used to project where future (sub)urban development is likely to occur. Third, this data is used in conjunction with our 0.5 m mapping product to estimate how much new residential lawn is likely to exist on these new developments. Finally, we discuss the potential impacts of this new lawn on water use in the region, as well as some of the strengths and weaknesses of the GEOMOD model.

We find that in 2005, 75.76 m^2 of lawn existed in the PIE study area. Under our Current Trends scenario, this increases to 81.04 (+6.96 %) square kilometers, while under Smart Growth is increases to 76.94 (+1.55 %) square kilometers (see Table 3). We estimate that today, lawn watering could be responsible for 879 million liters of water use. We project this could increase to 892 million liters in the Smart Growth scenario, and to 940 million liters under Current Trends.

A large number of research papers have been published in recent years based on the two-part premise that lawns are a major part of U.S. urban growth patterns, and that this land cover is associated with intensive resource use (primarily water and nitrogen inputs). This chapter presents the first research that produces large-extent, high-resolution tabulations of historic (2005) and future (2030, under two scenarios) lawn cover. Armed with such information, future researchers can now translate other data on resource use patterns to make statements about likely ecological implications associated with suburban lawn care, thereby testing the hypothesis that U.S. lawn care presents a significant environmental risk in suburbanizing watersheds.

Building on this, one fruitful future research avenue would be to increase efforts to acquire data on resource use patterns. Information on how much water is used, when, or why, by homeowners is simply not detectable using satellite-derived data. Rather, interviews and surveys are required for this purpose. Moreover, even though our 1,150 km^2 area represents one of the most extensive high resolution lawn studies to date, the rapid increases in inexpensive computing power and the availability of high-resolution remotely sensed data suggests that such mapping projects may soon be within reach of many researchers. As such, in the coming decade, we may be able to make statements about future land cover and associated resource use at state or even continental scales. Accordingly, another fruitful future research avenue is methods to accelerate the quantification of lawn area from remotely sensed imagery and the data management techniques to support such activities.

Acknowledgements The United States' National Science Foundation (NSF) supported this work via the following programs: Long Term Ecological Research via grants OCE-0423565 and OCE-1026859 for Plum Island Ecosystems and OCE-0620959 for Georgia Coastal Ecosystems, Coupled Natural Human Systems via grant BCS-0709685, Research Experiences for Undergraduates Site via grant SES-0849985, Decision-Making Under Uncertainty via grant SES-0951366, Urban Long Term Research Areas via grant BCS-0948984, and a supplement grant entitled "Maps and Locals (MALS)" via grant DEB-0620579. The work has also benefited from the NICHD funded University of Colorado Population Center (grant R21 HD51146) for research, administrative, and computing support. Any opinions, findings, conclusions, or recommendation expressed in this paper are those of the authors and do not necessarily reflect those of the funders.

References

Agarwal, C., Green, G., Grove, J., Evans, T. P., & Schweik, C. (2002). *A review and assessment of land-use change models: dynamics of space, time, and human choice, I*. University, ed. Bloomington: Centre for the Study of Institutions Population and Environmental Change.

Benz, U., Hofmann, P., Willhauck, G., Lingenfelder, I., & Heynen, M. (2004). Multi-resolution, object-oriented fuzzy analysis of remote sensing data for GIS-ready information. *ISPRS Journal of Photogrammetry and Remote Sensing, 58*(3–4), 239–258.

Berry, M. W., Flamm, R. O., Hazen, B. C., & MacIntyre, R. L. (1994). *The Land-use Change Analysis System (LUCAS) for evaluating landscape management decisions*. Knoxville: University of Tennessee.

Blaschke, T., & Strobl, J. (2001). What's wrong with pixels? Some recent developments interfacing remote sensing and GIS. *Interfacing Remote Sensing and GIS, 6*, 12–17.

Blaschke, T. (2010). Object based image analysis for remote sensing. *ISPRS Journal of Photogrammetry and Remote Sensing, 65*, 2–16.

Blaschke, T., Lang, S., Lorup, E., Strobl, J., & Zeil, P. (2000). Object-oriented image processing in an integrated GIS/remote sensing environment and perspectives for environmental applications. *Environmental Information for Planning, Politics and the Public, 2*, 555–570.

Burnett, C., & Blaschke, T. (2003). A multi-scale segmentation/object relationship modeling methodology for landscape analysis. *Ecological Modelling, 168*, 233–249.

Carrico, A. R., Fraser, J., & Bazuin, J. T. (2012). Green with envy: Psychological and social predictors of lawn fertilizer application. *Environment and Behavior, 45*, 427–454.

Castella, J., Trung, T., & Boissau, S. (2005). Participatory simulation of land-use changes in the northern mountains of Vietnam: The combined use of an agent-based model, a role-playing game, and a geographic information system. *Ecology and Society, 10*, 27.

Chow, W. T. L., Brennan, D., & Brazel, A. J. (2011). Urban heat island research in Phoenix, Arizona: Theoretical contributions and policy applications. *Bulletin of the American Meteorological Society, 93*, 517–530.

Domene, E., & Saurí, D. (2006). Urbanisation and water consumption: Influencing factors in the metropolitan region of Barcelona. *Urban Studies, 43*, 1605.

Domene, E., Saurí, D., & Parés, M. (2005). Urbanization and sustainable resource use: The case of garden watering in the Metropolitan Region of Barcelona. *Urban Geography, 26*, 520–535.

Engelen, G., White, R., & Nijs, T. (2003). Environment explorer: Spatial support system for the integrated assessment of socio-economic and environmental policies in the Netherlands. *Integrated Assessment, 4*, 97–105.

Fissore, C., Baker, L., Hobbie, S., King, J., McFadden, J., Nelson, K., & Jakobsdottir, I. (2011). Carbon, nitrogen, and phosphorus fluxes in household ecosystems in the Minneapolis-Saint Paul, Minnesota, urban region. *Ecological Applications, 21*, 619–639.

Giner, N. M., Polsky, C., Pontius, R. G., Jr., & Runfola, D. M. (2013). Understanding the determinants of lawn landscapes: A fine-resolution spatial statistical analysis in suburban Boston, Massachusetts, USA. *Landscape and Urban Planning, 111*, 25–33.

Glennon, R. J. (2002). *Water follies: Groundwater pumping and the fate of America's fresh waters*. Washington, D.C.: Island Press.

Glennon, R. J. (2009). *Unquenchable: America's water crisis and what to do about it*. Washington, D.C.: Island Press.

Gober, P., Brazel, A., Quay, R., Myint, S., Grossman-Clarke, S., Miller, A., & Rossi, S. (2010). Using watered landscapes to manipulate urban heat island effects: How much water will it take to cool Phoenix? *Journal of the American Planning Association, 76*, 109–121.

Groffman, P. M., Law, N. L., Belt, K. T., Band, L. E., & Fisher, G. T. (2004). Nitrogen fluxes and retention in urban watershed ecosystems. *Ecosystems, 7*, 393–403.

Hanlon, B., Short, J. R., & Vicino, T. J. (2010). *Cities and suburbs: New metropolitan realities in the US*. New York, NY: Routledge.

Harris, E. M., Martin, D. G., Polsky, C., Denhardt, L., & Nehring, A. (2012a). The role of emotions in suburban yard management practices. *Professional Geographer, 65*, 345–361.

Harris, E. M., Polsky, C., Larson, K., Garvoille, R., Martin, D. G., Brumand, J., Ogden, L. (2012b). Explaining U.S. lawncare practices to improve management: Evidence from suburban Boston, Miami, and Phoenix. *Human Ecology,* online.

Hilferink, M., & Rietveld, P. (1998). Land use scanner: An integrated GIS based model for long term projections of land use in urban and rural areas. *Journal of Geographical Systems, 1*, 155–177.

Hill, T. D., & Polsky, C. (2007). Suburbanization and drought: A mixed methods vulnerability assessment in rainy Massachusetts. *Environmental Hazards, 7*(4), 291–301.

House-Peters, L. A., & Chang, H. (2011). Urban water demand modeling: Review of concepts, methods, and organizing principles. *Water Resources Research, 47*, W05401.

Ipswich Utilities. (2011). Available from: http://www.ipswichutilities.org/

Isaaks, E. H., & Srivastava, R. M. (1989). *An introduction to applied geostatistics*. New York: Oxford University.

Johnston, R. A., Shabazian, D. R., & Gao, S. (2003). UPlan: A versatile urban growth model for transportation planning. *Transportation Research Record: Journal of the Transportation Research Board, 1831*, 202–209.

Karl, T., Fall, R., Jordan, A., & Lindinger, W. (2001). On-line analysis of reactive VOCs from urban lawn mowing. *Environmental Science & Technology, 35*, 2926–2931.

Klosterman, R. E. (1999). The what if? Collaborative planning support system. *Environment and Planning B, 26*, 393–408.

Krahe, J., Runfola, D. M., & Polsky, C. (2012). The impact of pricing structure on residential and seasonal water consumption in suburban Boston, MA, in Department of Economics. Worcester.

Krass, B. (2003). Combating urban sprawl in Massachusetts: Reforming the zoning act through legal challenges. *Environmental Affairs, 30*, 605–639.

Landis, J., & Zhang, M. (1998). The second generation of the California urban futures model. *Environment and Planning B: Planning and Design, 25*, 795–824.

Lang, R. E., Blakely, E. J., & Gough, M. Z. (2005). Keys to the new metropolis: America's big, fast-growing suburban counties. *Journal of the American Planning Association, 71*(4), 381–391.

Larson, K. L., Casagrande, D., Harlan, S. L., & Yabiku, S. T. (2009). Residents' yard choices and rationales in a desert city: Social priorities, ecological impacts, and decision tradeoffs. *Environmental Management, 44*, 921–937.

Lo, C., & Yang, X. (2002). Drivers of land-use/land-cover changes and dynamic modeling for the Atlanta, Georgia metropolitan area. *PE & RS- Photogrammetric Engineering & Remote Sensing, 68*(10), 1073–1082.

MAPC. (2010). *Metropolitan area planning council*. Boston: Metropolitan Area Planning Council.

MassGIS. (2011). Office of Geographic and Environmental Information. Commonwealth of Massachusetts, Executive Office of Energy and Environmental Affairs.

Mayer, P., & DeOreo, W. (1999). *Residential end uses of water*. Denver: American Water Works Association.

Milesi, C., Running, S. W., Elvidge, C. D., Dietz, J. B., Tuttle, B. T., & Nemani, R. R. (2005). Mapping and modeling the biogeochemical cycling of turf grasses in the United States. *Environmental Management, 36*, 426–438.

Milesi, C., Elvidge, C., & Nemani, R. (2009). Assessing the extent of urban irrigated areas in the United States. In *Remote sensing of global croplands for food security*. San Fransisco: NASA Ames Ecological Forecasting Lab. Retrieved from: http://ecocast.arc.nasa.gov/pubs/pdfs/2009/Milesi_Urban_BookChapter.pdf

Nielson, L., & Smith, C. L. (2005). Influence on residential yard care and water quality: Tualatin watershed, Oregon. *JAWRA Journal of the American Water Resources Association, 41*, 93–106.

Pijanowski, B., Gage, S., & Long, D. (1997, February). *The land transformation model.* Paper presented at: Land Use Modeling Workshop, Sioux Falls.
Polsky, C., Pontius Jr., R. G., Decatur, A., Giner, N., Rahul, R., & Runfola, D. M. (2012). *Mapping lawns using an object-oriented methodology with high-resolution four-band aerial photography: The twenty-sex towns of the Ipswich and Parker River Watersheds, Massachusetts.* In George Perkins Marsh Working Paper. Worcester: Clark University.
Pinto, P., Cabral, P., Caetano, M., & Alves, M. F. (2009). Urban growth on coastal erosion vulnerable stretches. *Journal of Coastal Research, 56*(2), 1567–1571.
Pontius, R. G. J., Cornell, J., & Hall, A. S. C. (2001). Modeling the spatial pattern of land-use change with GEOMOD2: Application and validation for Costa Rica. *Agriculture, Ecosystems & Environment, 85*, 191–203.
Priest, M., Williams, D., & Bridgman, H. (2000). Emissions from in-use lawn-mowers in Australia. *Atmospheric Environment, 34*, 657–664.
Robbins, P. (2007). *Lawn people: How grasses, weeds, and chemicals make us who we are.* Philadelphia: Temple University Press.
Robbins, P., & Birkenholtz, T. (2003). Turfgrass revolution: Measuring the expansion of the American lawn. *Land Use Policy, 20*, 181–194.
Robbins, P., & Sharp, J. T. (2003). Producing and consuming chemicals: The moral economy of the American lawn. *Economic Geography, 79*, 425–451.
Roy Chowdhury, R., Larson, K., Grove, M., Polsky, C., Cook, E., Onsted, J., & Ogden, L. (2011). A multi-scalar approach to theorizing socio-ecological dynamics of urban residential landscapes. *Cities and the Environment (CATE), 4*, 6.
Runfola, D. M., Polsky, C., Nicolson, C., Giner, N., Pontius, R. G., Jr., Krahe, J., & Decatur, A. (2013). A growing concern? Examining the influence of lawn size on residential water use in suburban Boston, MA, USA. *Landscape and Urban Planning, 119*, 112–123.
Runfola, D. M., & Pontius, R. G., Jr. (2013). Quantifying the temporal instability of land change transitions. *International Journal of GIS, 27*(9), 1696–1716.
Silva, E. A., & Clarke, K. C. (2002). Calibration of the SLEUTH urban growth model for Lisbon and Porto, Portugal. *Computers, Environment and Urban Systems, 26*, 525–552.
Tobler, W. R. (1970). A computer movie simulating urban growth in the Detroit region. *Economic Geography, 46*, 234–240.
Tu, J., Xiz, Z., Clark, K. C., & Frei, A. (2007). Impact of urban sprawl on water quality in Eastern Massachusetts, USA. *Environmental Management, 40*, 183–200.
United States Census Bureau (2011). Decennial Census. http://www.census.gov/.
US Census (2013). http://www.census.gov/
Veldkamp, A., & Fresco, L. (1996). CLUE-CR: An integrated multi-scale model to simulate land use change scenarios in Costa Rica. *Ecological Modelling, 91*, 231–248.
Waddell, P. (2002). Modeling urban development for land use, transportation, and environmental planning. *Journal of the American Planning Association, 68*, 297–314.
Walker, R., Drzyzga, S. A., Li, Y., Qi, J., Caldas, M., Arima, E., & Vergara, D. (2004). A behavioral model of landscape change in the Amazon basin: The colonist case. *Ecological Applications, 14*, 299–312.
Wentz, E. A., & Gober, P. (2007). Determinants of small-area water consumption for the city of Phoenix, Arizona. *Water Resources Management, 21*, 1849–1863.
Yang, X., & Lo, C. (2003). Modelling urban growth and landscape changes in the Atlanta metropolitan area. *International Journal of Geographical Information Science, 17*, 463–488.
Zhou, W., Troy, A., & Grove, M. (2008). Object-based land cover classification and change analysis in the Baltimore metropolitan area using multitemporal high resolution remote sensing data. *Sensors, 8*, 1613–1636.

Land Use-Land Cover Dynamics at the Metropolitan Fringe

Daniel Felsenstein, Michal Lichter, Eyal Ashbel, and A. Yair Grinberger

Abstract Diverse pressures for change operate at the outer metropolitan fringe. This paper examines the spatial and temporal dynamics of change in this area. We set up a simple model that incorporates spatial and temporal dynamics of functional (land use) and structural (land cover) interactions. We posit that land use (development) changes the ecosystem functions at the edge of urban areas expressed in change in land cover. Additionally, the characteristics of land cover (forest, agriculture, bare soil, neighboring cover etc.) mutually influence the land use. We estimate a model where land values and land use are jointly determined while land use and land cover interact recursively. We use historical data, probability estimation and land use simulation to generate panel data of future patterns of land value, land use and land cover at the outer edge of the Tel Aviv metropolitan area for the period 1995–2023. The modeling system combines panel 2SLS (2-stage least squares) estimation to investigate land value-land use interactions. Land use-land cover dynamics are estimated using panel MNL (multi-nomial logit) estimation. Results of simple simulations of the probability of land cover change are presented. When coupled with an appropriate biodiversity model, this system could potentially be extended to forecasting other aspects of the environmental stress of metropolitan expansion, for example impacts on vegetation or ecological dynamics.

1 Introduction

Worldwide, the metropolitan fringe is under heavy development pressure and in a constant state of flux. It interfaces between the built and the natural environments and serves as the battleground on which land use conflicts are fought.

D. Felsenstein (✉) • M. Lichter • E. Ashbel • A.Y. Grinberger
Department of Geography, Hebrew University of Jerusalem, Mt Scopus, 91905 Jerusalem, Israel
e-mail: msdfels@mscc.huji.ac.il; mlichter@gmail.com; eyal.ashbel@gmail.com; solomonico@gmail.com

D. Malkinson et al. (eds.), *Modeling of Land-Use and Ecological Dynamics*, Cities and Nature, DOI 10.1007/978-3-642-40199-2_8, © Springer-Verlag Berlin Heidelberg 2013

Unsurprisingly, much of the effort in the area of modeling urban development is focused on land use change in this area (Ravetz et al. 2013). Invariably, this interest centers on the way in which human agents (developers, households, firms, governments etc.) generate a market for land. Under conditions of supply and demand, a price structure emerges and the market clears at the metropolitan fringe. Human behavior creates land uses of different kinds and these further impact on the emergence of a price structure that re-impacts on human behavior. In this loop, land prices are created endogenously (Felsenstein and Ashbel 2010).

This picture however is only partial. This is because land use is intricately connected to land cover. For example, high density development (land use) may be related to a particular land cover such as bare soil, low residential development may be associated with forest land or natural vegetation and so on. Once we extend the casual loop from the effect of human behavior on land use to its effect on land cover, we are essentially looking at the way in which anthropogenic shocks impact biophysical and ecological processes. This extended loop however is also circular. Land cover such as forest land, plantations and orchards also make for the attractiveness of areas adding to their economic value. In addition it serves an ecological purpose that also has monetary value for example by removing air pollutants, mitigating micro-climates, absorbing rainfall and reducing run-off.

This paper explores these interactions in the context of the metropolitan fringe. We extend the human behavior-land use causality to the realm of land cover and look at the dynamic feedback links between them. Based on these connections, we first estimate the joint determination of land values and land use. Then, we progress to tracing the temporal and spatial impacts of anthropogenic disturbances on land cover at the urban edge, through the mediation of land use. We do not attempt to forecast land use or land cover change. Rather, we take these predictions as inputs for a model that causally traces the interactions between land values, use and cover.

The driver of this process of market formation is land values but in addition we add temporal dynamics in which agents of change learn over time and spatial dynamics in which agents affect their neighbors. Current land use modeling does not generally specify the role of land cover and is only recently showing an interest in spatial and temporal dynamics (Irwin 2010). The proposed model is grounded in human behavior and can be potentially extended to incorporate biophysical processes. We only look at land cover conversion and not modification (Alberti and Waddell 2000). Conversion refers to change form one land cover to another, for example, form forest cover to urbanized cover. Modification describes the process of changing conditions within a given cover type for example, from deciduous forest to evergreen. We focus on land conversion as this represents the most direct and stark change taking place at the metropolitan fringe. Our approach allows for simulations of change in land values, residential densities and land use in the metropolitan fringe and their ripple-through effects over time and space. While we only observe intra-metropolitan fringe interactions, we believe that the modest extension to land cover presented here has potential for generating new insights into the dynamics of urban development. For example, questions such as 'do high residential densities tend to develop on particular land surfaces?', 'does urban development over time tend to seek out better quality land as supply tightens?' and 'do agents of change tend to affect each other?' are all issues high on both the policy and praxis agendas.

2 Literature Review: Land Use-Land Cover Modeling

The complexity of land use-land cover interactions has led to many different analytical approaches to modeling these relationships. These range from statistical-econometric, through analytic (equilibrium-based) to simulation- driven (micro, cellular agent and multi-agent variants). The differences between these different modeling systems has been comprehensively reviewed elsewhere (Irwin 2010; Parker et al. 2003; van Schrojenstein Lantman et al. 2011). Perhaps the most salient issue in studying the land use-land cover nexus is the explicit treatment of feedback mechanisms inherent in this connection. Verburg (2006) has addressed this issue by outlining the most common forms of feedback. However, the notion becomes all the more important when looking at dynamics. A typical feedback mechanism links social and biophysical systems. Thus, land use (an anthropogenic system) can degrade land cover (a biophysical system) causing a change in future land use. In the dynamic variant of this feedback earlier land use (t − 1) affects contemporaneous land cover (t) which in turn affects future land use (t + 1). Additionally, a second feedback can occur across space (Verburg 2006). This generally involves land use change at one scale (for example regional demand for housing) impacting of land cover at another scale (the uprooting of local orchards). Again, in a spatially dynamic setting the feedbacks need to be modeled as spatial dependence in that local land cover might be affected by regional land use but also by neighboring land use/cover (Verburg et al. 2004).

Typically, dynamic feedbacks can be incorporated by using dynamic simulation models. The dynamics of land use- land cover change are often modeled within the framework of existing micro- simulation models. In this context the UrbanSim model (Waddell and Ulfarson 2003) provides particularly fertile ground. Urban development through the agency of households firms and developers provides the key socio-economic pathway to change that finds expression in land use. Micro-simulation models are particularly useful at generating patterns of land use change at a high level of spatial resolution and at providing estimates of other socio-economic activities associated with land use change (such as change in population composition, land values etc.). When this framework is combined with patterns of land cover change, it takes but a small extension to estimate the probabilities of land cover transitions.

Empirical studies have shown the inter-relationships between particular land uses and land cover types, for example multi-family residences and impervious surfaces and single family residences and forest cover (Alberti 2005, 2008). The extension to land cover raises a series of issues relating to unchartered territory of urban development and ecological outcomes. This is because the biophysical properties of change to land cover have a host of tertiary effects on soil quality, land productivity, run off, sedimentation rates etc. These are not always considered as part of the anthropogenic footprint of urban development that is heavily focused on land consumption infrastructure congestion and pollution.

The land use-land cover relationship has been extended to incorporate reciprocal influences on the ecological environment. Alberti and Waddell (2000) sketch out how the UrbanSim land use forecasting system can be coupled with a biophysical model to estimate the impact of land cover change on hydrology and nutrient cycling. An empirical application of this integration relating to avian diversity is reported by Hepinstall et al. (2008). They use land use inputs from the UrbanSim model and land cover change simulations from Landsat data in order to predict changes in the composition of avian species in Washington State. Their work provides one of the few empirically operationalized models that takes land use-land cover models to the realm of ecological response within an integrated framework.

Dynamic simulation is undoubtedly pushing forward the frontiers of interdisciplinary work to understand the feedbacks between physical, economic and ecological change. Coupled with high resolution spatial and temporal data, efforts are no longer solely aimed at understanding landscape change as the outcome of a collective response to attain the highest and best use from land given bio-physical constraints. Rather, recent modeling tends to stress ‘neighborhood’ influences, idiosyncratic preferences, regulation and the influence of time. Together these make for a much more challenging modeling environment.

3 A Conceptual Model

We posit a particular causal sequence in the interactions between land use, land cover and human agency (Fig. 1). We use a bottom-up approach (Irwin et al. 2009) in which anthropogenic influences are assumed to generate a market for land and generate change in land use through the utility-maximizing activities of households, firms, land developers and policy makers. This in turn has a mutual effect on land values. The choice to convert a parcel of land to a different use or intensify current use is jointly determined with the expected return from that parcel. This simultaneity between land values and property prices on the one hand and other features of the land market such as regulatory practices (Ihlanfeldt 2007) and new construction (Mayer and Somerville 2000) is the subject of much attention in the urban economics and real estate literature. Land use and land cover are also posited to be causally related. We expect interactions to be recursive rather than simultaneous as temporal and spatial dynamics play a more pronounced role. Land use and land cover conversion are lengthy processes that do not react instantly to market signals (Verburg et al. 2004; Verburg 2006) and we thus assume a recursive process. The framework outlined in Fig. 1 can potentially be extended to incorporate more explicit ecological and biophysical inputs. If coupled with a biodiversity model

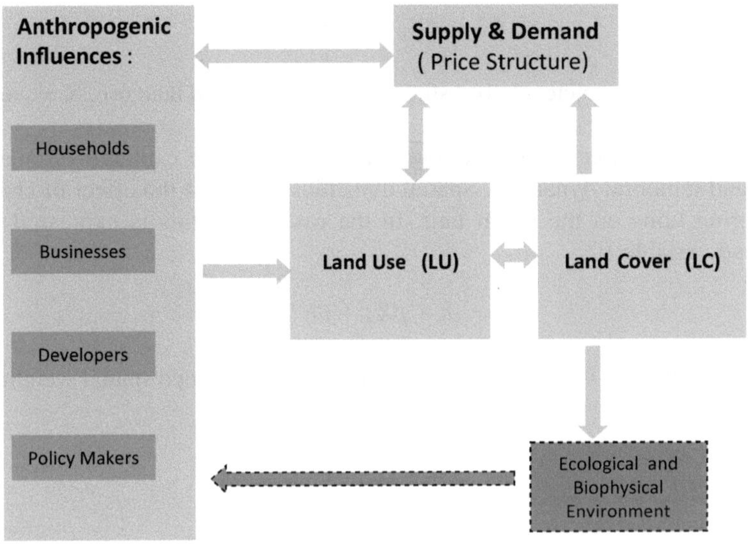

Fig. 1 Conceptual model

for example, the system would be capable of forecasting the environmental stress of metropolitan expansion on vegetation and ecological dynamics.

In line with classic models of urban structure, land value is expected to be directly related to residential density and gross number of units. It is expected to be inversely related to distance from the CBD and probably directly related to highway accessibility. Potential endogeneity exists of course between land values and residential density used here to represent land use. We expect land values to have a direct and positive effect on intensity of land use. In addition, temporally and spatially lagged residential densities are expected to have a similar effect on current land use density. Commercial activity is also posited to be directly related to residential land use intensity. Conversely, distance to the metropolitan center and to highways is expected to reduce pressure on residential density.

Land cover change consists of a set of spatially explicit choices relating to site-based land cover transitions. The likelihood of a single unit (grid cell) changing from one discrete land cover class to another class is a function of temporally lagged land cover, land use and neighboring land use, that represent endogenous pressure and endogenous attributes of the unit such as distance from the metropolitan center or distance from central points of accessibility such as highways.

Land value (V) land use (U) in unit i and time t are co-determined (Felsenstein and Ashbel 2010) such that:

$$V_{it} = \lambda U'_{it} + \beta_{it}\boldsymbol{X}_{it} + \varepsilon_{it} \tag{1}$$

$$U_{it}^{'} = \delta V_{it} + \beta_{it} \boldsymbol{X}_{it} + v_{it} \tag{2}$$

where: λ, β, δ = parameters to be estimated, U' = estimated land use, X = vector of cell (unit) attributes.

Interactions between land use and land cover (C) are captured through their spatial and temporal dynamics. Spatial dynamics represent the effect of change in neighboring units on the target unit. In the case of U, this is expressed by the asterisked variable U_{it}^{*},

$$U_{it} = \alpha_i + \beta X_{it} + \theta U_{it}^{*} + u_{it} \tag{3}$$

where the spatial lag is generated by appropriate row-summed spatial weights (w_{ij}).

$$U_{it}^{*} = \sum_{j \neq i}^{N} w_{ij} U_{jt}$$

Temporal dynamics are expressed as the time lagged value of the target unit,

$$U_{it} = \alpha' + \beta' X_{it} + \theta' U_{it-1} + u_{it} \tag{4}$$

While U is often measured as a discrete variable, we follow the tradition of observing land use as a continuous variable represented by the intensity of use of artificial surfaces (Alberti 2005; Lopez et al. 2001; Ravetz et al. 2013). Operationalizing U and C interactions we posit that U (residential density) is determined by V, previous residential density and neighboring density (i.e. spatially and temporally lagged U), the presence of commercial activity and various distance attributes (K), as follows:

$$U_{it} = \alpha U_{it-1} + \lambda \sum_{i \neq n}^{N} w U_{it} + \delta V_{it} + K X_{it} + v_{it} \tag{5}$$

Land cover (C) change is subsequently expected to be influenced by previous land cover, current (U^{*}) and temporally lagged land use and a vector of distance measures:

$$C_{it} = \gamma C_{it-1} + \alpha U_{it-1} + \lambda U^{*}{}_{it} + K X_{it} + u_{it} \tag{6}$$

With this system set up, scenario based simulations can be derived. Impulse response movements can be simulated to answer question such as how would an increase in current land value affect the likelihood of land cover change from bare soil to built area in the future, i.e. $\partial C_{t+1} / \partial V_t$? The matrix formulation and mathematical structure of the recursive estimation are presented in Appendix 1.

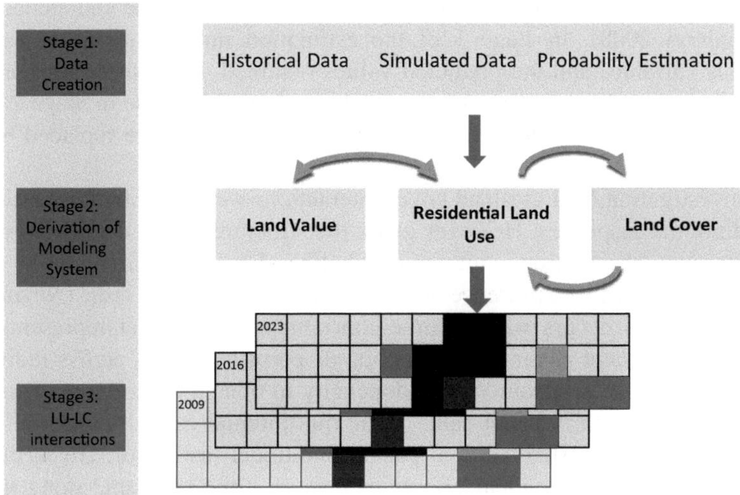

Fig. 2 Method

4 Estimation Strategy

The estimation strategy is outlined in Fig. 2. As the key issue here is the interaction between temporal and spatial dynamics and not land use/cover forecasting, we use opportunistic mix of historical, simulation and probability generated data on which to test our method. The nature of the data generation process is described below. Our main thrust is focused on deriving the estimation system and then generating some simple statistical simulations highlighting the interplay between land values, residential density (representing land use) and land cover. We are cognizant of the reservations associated with both econometric land use modeling coupled with spatial simulation (endogeneity) and raster GIS-based models (spatial dependence) (Irwin 2010). The estimation strategy adopted addresses these issues directly.

Our approach is based on the joint determination of land values and (residential) land use. This simultaneity raises the specter of endogeneity i.e. the notion that one or more of the explanatory variables of land value and use is jointly determined with the dependent variable, typically through an equilibrium mechanism. Simultaneity is a common issue in economic behavior. A key feature of simultaneous equations models such as supply–demand systems is that a change in either supply or demand will affect both the equilibrium price and quantity in the market, so that by construction both variables are correlated with any shock to the system. An explanatory variable that is determined simultaneously with the dependent variable is generally correlated with the error term. OLS estimation of such a model will produce biased and inconsistent estimators. In simultaneous equation systems such as ours, the common strategy is to identify the parameters using instrument variables (IV) and to estimate using two stage least squares (2SLS). As we use panel data, the panel 2SLS variant calls for first, eliminating unobserved effects by using first differencing and second, using IV's for the endogenous variables (Baltagi 2008). In stage 1 of the estimation models are fitted using all exogenous variables and the predicted values obtained. From these reduced-form estimates, predicted values from each model are obtained for use in Stage 2. In this stage the original endogenous variables from the first stage are replaced by their fitted values.

For investigating land use-land cover interactions we use probability estimation for multinomial responses. However given the structure of our data (a short panel data with many observations), maximum likelihood estimators may not be consistent or efficient due to the presence of the 'incidental parameter' issue (Neyman and Scott 1948). This occurs when a time-constant parameter that represents some innate feature of land cover (such as slope or porosity) varies across individuals units and may serve as a source of endogeneity in other x_{it} regressors. This limits probability estimation in panel data. In the multinomial case, Baltagi (2008) and Lee (2013) have suggested using a 'panel conditional' estimator. Given the existence of unobserved individual effects α_i that are time constant, along with X_{it} observed characteristics between individuals and over time, we follow Haan and Uhlendorff (2006) using a STATA routine for estimating an MNL model for land cover using panel data. In the regular MNL approach, the probability of making choice j conditional on observed characteristics X_{it} and unobserved individual effects α_i is:

$$\Pr(j|X_{it}\alpha_i) = \frac{\exp(X_{it}\beta_j + \alpha_{ij})}{\sum_{k=1}^{J} \exp(X_{it}\beta_j + \alpha_{ij})} \tag{7}$$

However as the choice probabilities are conditioned on α_i, this requires integrating Eq. 1 on the distribution to get the likelihood (L) as follows:

$$L = \prod_{i=1}^{N} \int_{-\infty}^{\infty} \prod_{t=1}^{T} \prod_{j=1}^{J} \left\{ \frac{\exp(X_{it}\beta_j + \alpha_{ij})}{\sum_{k=1}^{J} \exp(X_{it}\beta_j + \alpha_{ij})} .. \right\}^{d_{ijt}} f(\alpha)d\alpha \tag{8}$$

where $d_{ijt} = 1$ if unit i transitions to alternative j at time t and 0 otherwise. The coefficient vector and the unobserved heterogeneity term of one category are set to 0 for identification of the model. We assume that unobserved heterogeneity α is identically and independently distributed over all units and follows a multivariate normal distribution with mean a and variance-covariance matrix $\mathbf{W}$,$\alpha \sim f(a,W)$.

The spatial nature of units (expected land cover in neighboring units) is considered a feature of X_{it}. On the one hand, this obviates the need for estimating an enormous spatial weight matrix given the grid cell structure of the data (see below). On the other hand, it calls for appropriate treatment of spatial dependence between observations. While the routine way of dealing with this is via a row-summed weight matrix (see Eq. 3 above), this is obviously untenable in our case. Consequently we use a dedicated Python script to create spatial lags by automatically averaging the values of the eight neighboring cells around the target cell. It should be noted that while this approach technically generates values for 'neighbors', the real spatial lag is still not properly identified due to the circular 'I-am-my-neighbor's-neighbor' property of spatial relations.

5 Data

5.1 Data Preparation

In line with other studies of land use land cover interactions (Alberti 2005; Hepinstall et al. 2008) we exploit various sources in order to assemble the land use and land cover data. Our data set cover the period 1995–2023, so by construction we are using synthetic data. Land cover data for historical periods come from a compilation of different GIS data covers: 1995 data comes from the National Outline Plan (NOP) #35, 2002 data from the central Bureau of Statistics (CBS) and 2009 data comes from the Survey of Israel. This spatial series is extended at 7 year lags through to 2023 using probability estimation as outlined in Appendix 2. This yields five basic land cover classes: urbanized (built area), bare soil, forests, orchards/plantations and fields. 'Urbanized areas' is a catch-all category that acts as the baseline and includes any grid cell where built structures or urbanized land cover such as parks and gardens form the majority footprint in the grid cell. For comparison, a remote sensing study of natural habitats in Israel used a similar schema based on bare soil, dunes, herbaceous trees, shrubs, agricultural cover and water bodies (Levin et al. 2011).

Land use relates to residential land use and is measured by residential intensity. Other metrics of land use that could have been used include land use heterogeneity and connectivity, but intensity is often used as the default choice (Alberti 2005). The source of this data and of all the other attributes of grid cells such as residential land value, residential units, non residential area etc. is the UrbanSim 3.0 model calibrated for the Tel Aviv metropolitan area (Felsenstein et al. 2007). The historical starting point for this data is 1995 and all subsequent periods relate to simulated data. The simulated data is generated at the 250 $\times$ 250 m grid cell level. Given this source, the data consists of five balanced panels each with nearly 10,000 grid cells. The lack of symmetry between the temporal ($T = 5$) and spatial ($N = \sim 10{,}000$) components of the data, should be noted. Despite this a-symmetry, the data was treated as a panel rather than pooled cross-section thereby utilizing all data fully. Obviously, this structure constrains the use of a spatial weight matrix and spatially lagged variables are constructed in a manner described above.

Table 1 describes the variables, their sources and the way they are constructed.

Table 1 Variables, method of construction and sources

Variable	Calculation method	Source
Land cover 1995	Raster to grid cells using zonal statistics	NOP/35
Land cover 2002	Vector to grid cells using zonal statistics	CBS
Land cover 2009	Vector to grid cells using zonal statistics	Survey of Israel
Residential density	Gross residential unit density: number of residential units divided by grid cell area	NOP/35
Residential units	Residential units allocated to grid cells with centroids within the census tract.	1995 National Census
Residential land value	Average value per sqm * residential footprint sqm	Israel Land Administration-results of land auctions
Nonresidential land value	Average value per sqm * non-residential footprint sqm	Israel Land Administration –results of land auctions
Non residential area	area of the grid cell * (1-fraction residential land)	NOP/35
Distance to 1 digit highway	Euclidean distance (sqm) from centroid of cell to highway	MAPA 2009
Distance to TLV metropolitan CBD	Euclidean distance (sqm) from centroid of cell to metropolitan core boundary (defined by CBS)	CBS

5.2 *Data Description*

The metropolitan fringe is a transitional area under urban influence but with rural morphology. We define this transitional zone as a 5 km buffer that starts at the outer boundary of the metropolitan area and extends inwards. The area contains some 154 municipal entities, 37 of them are urban municipalities and the rest, small communities and villages. The population of the fringe zone is 670,000 and average household income is nearly 18,000 sh (2008). The Tel Aviv metropolitan area (minus the urban fringe area) has a population of 2,630,000 and average household income similar to the fringe area.

Table 2 describes the main changes in key variables in the metropolitan fringe over the period 1995–2009 and compares these with corresponding changes in the metropolitan area. Most of the pressure for change stems from residential expansion in five growth points. The city of Ashdod increased its population by 63 % from 126,000 in 1995 to 206,500 in 2009. Perhaps the most dramatic growth occurred in the area of Modiin where a new city was constructed and populated virtually ex nihilo over this period resulting in population growing from 10,600 in 1995 to 72,700 in 2009. The small town of Rosh Ha'ayin saw a 30 % population increase from 29,000 to 38,000, the trio of small Arab towns (Tira, Taibe and Kalansua) experienced population growth of 40 % (from 54,000 to 76,700) and the medium sized city of Kfar Saba saw the lowest rate of growth (18 %) with population expanding from 70,500 to 83,600. In all, these 5 growth poles account for over 70 % of the entire population in the metropolitan fringe.

Table 2 Land use and land cover changes in the metropolitan fringe, 1995 and 2009

	1995		2009	
	Metro area	Fringe	Metro area	Fringe
Residential land value[a]	130,624	53,907	3,402,274	3,167,253
Residential units[b]	679.1	100.1	868.3	207.5
Non-resid land value[c]	3,896,467	3,369,864	5,750,738	5,606,086
Commercial area[d]	3.503	1.334	24.546	14.921
Land cover[e]				
Bare soil	134.4	140.9	129.8	210.0
Forests	40.1	34.5	56.8	48.9
Orchards	141.5	93.3	96.8	62.1
Fields	143.3	174.2	183.4	131.9
Urbanized area	376.5	119.1	424.6	165.1

[a]Average value per grid cell (Israeli Shekels m, current prices)
[b]Total residential units (th)
[c]Average value per grid cell, commercial, industrial and governmental land uses (Israeli Shekels m, current prices)
[d]Total footprint area, m sq.m.
[e]Total area, m sq.m.

The data in Table 2 reflect these dramatic changes. The gaps between the metropolitan area and the fringe in residential land values and number of units in 1995 are closed or equalized by 2009. Non residential land values are roughly similar in both periods whereas the share of increase in commercial areas (assumed to follow the growth of residential areas) is particularly pronounced in the fringe. In terms of land cover change, there is a dramatic increase in bare soil area in the fringe, often a precursor of land conversion. In the metropolitan area the amount of bare soil actually contracts. Forest cover also increases in both areas. However as this is generally due to public initiative it could be a knee-jerk reaction to development pressures in other parts of the fringe and an attempt to curtail development.

Table 3 highlights the urban fringe alone and presents simulated data out till 2023. The main feature of these data is the expected entrenchment of patterns started over the period 1995–2009. Residential units, densities and land values are expected to continue to grow unabated and to generate large increase in attendant commercial areas. With respect to land cover, forests (after an initial growth period) orchards and fields are all expected to contract. In contrast, the prevalence of bare soil (as an interim land cover) continues to expand and along with it urbanized land cover (buildings and other non-natural covers) is expected to increase.

Figures 3 and 4 depict the main changes expected in land use and land cover over the simulation period. For land use, we see the progression of increasing colonization and intensification over the period as residential densities increase in both scale and scope. Figure 3 highlights the example of Modiin where public sector-led development has initiated a process in the south east corner of the fringe area. The intensity of development is not forecast to rival that of the more established urban centers in the fringe but the scale of low intensity development is large.

Table 3 Land values, use and cover in the metropolitan fringe, simulated data

	1995	2002	2009	2016	2023
Residential land value[a]	53.9	2,224.5	3,166.8	4,172.8	4,768.8
Residential units[b]	100.1	158.2	207.5	271.3	324.3
Residential density[c]	162	256	336	439	524
Commercial area[d]	1.334	8.918	14.921	22.851	28.956
Land cover[e]					
Bare soil	140.9	161.7	210.0	235.4	260.1
Forests	34.5	74.8	48.9	36.5	23.5
Orchards	93.3	92.0	62.1	57.5	50.5
Fields	174.2	183.4	131.9	119.1	91.2
Urbanized area	119.1	106.1	165.1	169.5	186.7

[a]Average value per grid cell (Israeli Shekels m, current prices)
[b]Total residential units (th)
[c]Average units per sq km
[d]Total footprint area, m sq.m.
[e]Total area, m sq.m

The land cover changes in Fig. 4 portray the spatial detail of the aggregate data in Table 3. The virtual disappearance of orchards and forests in the urban fringe by 2023 could not be discerned from data relating to the earlier periods (Fig. 4). Again focusing on the case of Modiin, we can note an upsurge of forest and field cover after the development of the area. This process has also been noted elsewhere (Levin et al. 2011) where initial urban encroachment into the fringe is characterized by an abrupt decrease in vegetative cover as land is stripped bare in preparation for urban expansion. After a period of time, the vegetation returns in the form of man-made land cover (parks, gardens and sculptured open spaces).

6 Results

6.1 Estimation Results

We employ statistical estimation and impulse response simulation. Equations 1, 5 and 6 are estimated as a system with spatial and temporal lags acting as the links. Impulse response shocks (simulations) are used to disturb the system and the perturbations are grounded in the spatial and temporal lag structures dictated by the model. Given the lattice structure of the spatial data, the results of the simulation are presented in map form.

The two-way causality and system feedbacks inherent in estimating Eqs. 1 and 2 above are a source of endogeneity. Land value is both driven by and drives residential density. We use panel 2SLS to deal with this and specify instrument variables for each stage of the regression. As we use fixed effects, the R^2 and constant in both the land value and the residential density models are very large

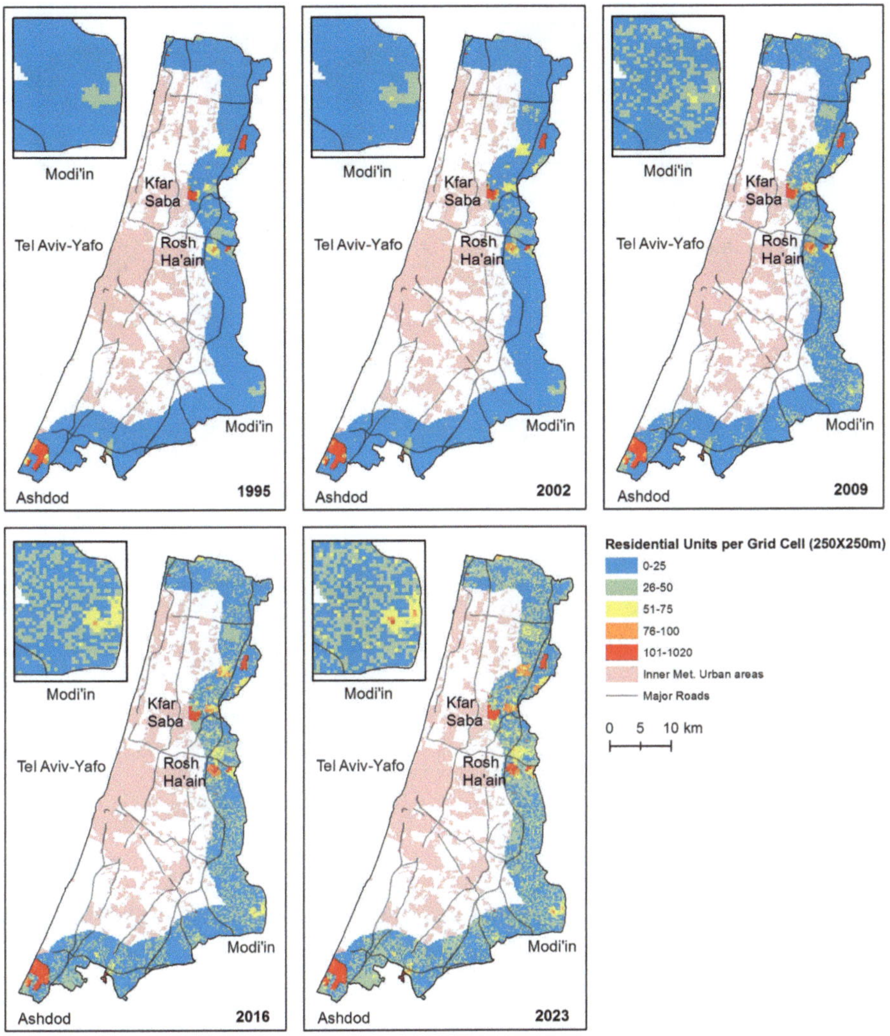

Fig. 3 Land use change 1995–2023 in the metropolitan fringe

(Table 4). The latter indicates that there is much going on in the model that is not explained by the regressors. In the land value model units have the right sign but density is counter intuitive. The effects of distance (both centrality and accessibility) are inversely related to value and significant. In the land use model, land value again displays the opposite sign to that expected. Spatial (RD neighbors) and temporal (RD lag) effects of land use have a direct and positive effect on current land use. Commercial activity is also directly related while the distance measures display inverse signs as expected and distance to the CBD (centrality) is not significant.

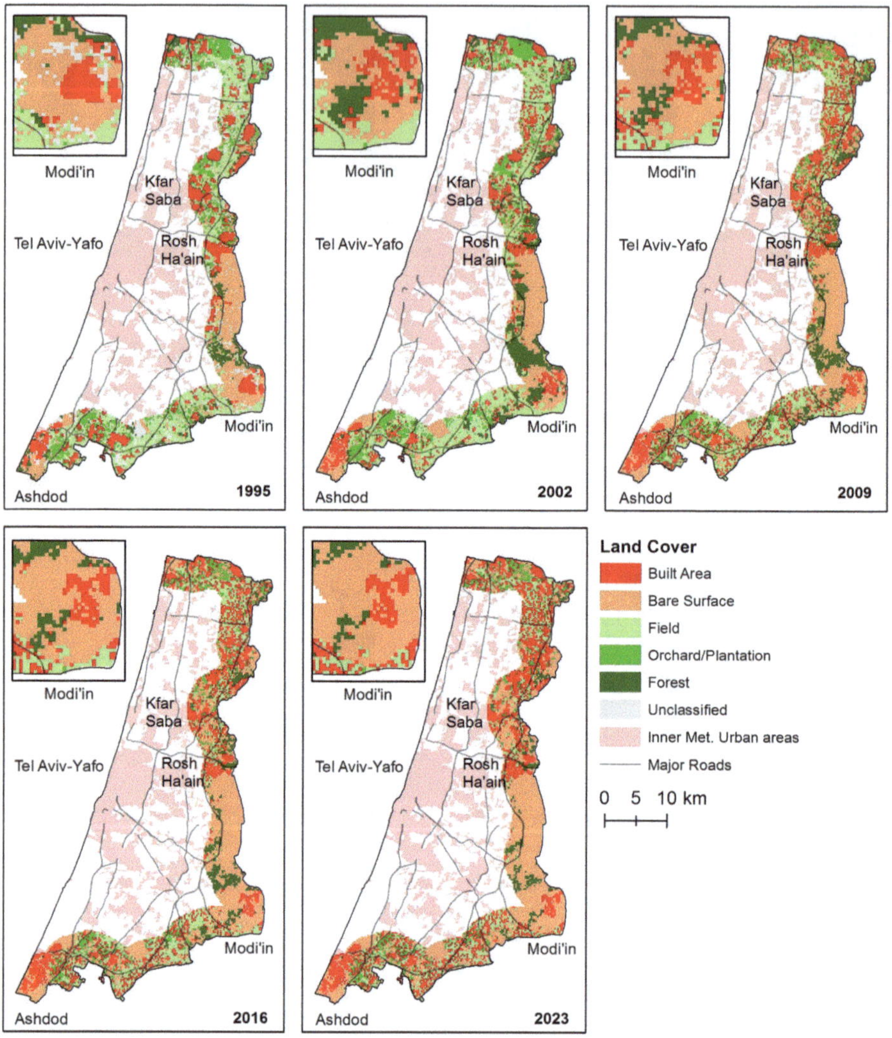

Fig. 4 Land cover change 1995–2023 in the metropolitan fringe

Table 5 shows the estimated results for Eqs. 5 and 6 using panel MNL regression. The recursivity in this estimation is through the current and lagged effect of land use. The coefficients in Table 5 need to be interpreted as the effect of a one unit increase in the regressor, on the relative log odds of the land cover (bare soil, forests etc.) transitioning to urbanized (the baseline category). For example, we can see that a unit change in residential density is associated with a 0.074 decrease in the probability of bare soil cover transitioning to urbanized. For three out of the four land covers an increase in residential density leads to a decrease in the chance of that land cover staying in its current state, but this result is only statistically

Table 4 Panel 2SLS estimation of land values and residential density (with fixed effects)

Land value		Residential density	
Constant	83.016	Constant	25.586
Resid density (RD)	−1.413	Land value	−0.706
Units	1.181	RD (lag)	−0.085
Dist. h'way	−0.012	RD (neighbors)	0.061
Dist CBD	−0.0007	Commercial area	0.215
		Dist h'way	−0.003
		Dist CBD	−0.00023*
Adj. R^2	0.944	Adj. R^2	0.946
F-statistic	86.105	F-statistic	71.58
SS resid	80,989.2	SS resid	12,003.3
Durbin Watson (DW) statistic	1.502	DW	1.469

All continuous variables in ln
Notes: all coefficients significant unless indicated (*)
IV specification Stage 1: lags, neighbors, squared values for variables: RD, units
IV specification Stage 2: lags, neighbors, squared values for variables: land value, units commercial area

significant for bare soil. Land cover in the earlier period also decreases the log odds of a current land cover staying as is, which implies that the fringe is in a state of flux. The same is true of change in residential density in an earlier period. Again, this decreases the likelihood of land cover staying in its present state. All told, changing accessibility or centrality also increases pressure for change, further underscoring the volatility of natural land covers in the urban fringe. It should be noted however while all the coefficients are significant they are not all consistently negative. This may indicate model instability.

6.2 *Simulation Results*

Given the above estimations, we use dynamic simulation for some general land cover change scenarios. This approach has shocks propagating across time (7 year lags) and space (grid cells). In all cases we observe the ratio of change between the baseline change and the scenario change. Due to the lack of symmetry between the spatial and temporal resolutions of this exercise, we expect most of the change to be driven by spatial dynamics. Because of the small cell size we do not shock individual spatial units but prefer to simulate a system wide change such as a rise in land values in time *t* and observe how this eventually affects the likelihood of land cover change having been mediated by land use and temporal and spatial effects on the way. To this end, we report general impulse responses from scenarios even though our time lags are very coarse. This of course precludes our ability to be able to trace impulse responses dying out over time. On the other hand, the 7 year time lag here reflects the time span over which land use and land cover change take effect given the constraints of the planning system. The scenarios are as follows:

Table 5 Probability of land cover change: MNL regression with panel data

	Land cover			
	Bare soil	Forests	Orchards, plantations	Open fields
Constant	12.595	7.646	8.891	5.931
Land cover – lag	−.3.116	−2.399	−1.937	−1.277
Resid. density	−.074	.025	−.008*	.010*
Resid. density – lag	.008*	−.126	−.045	−.135
Distance h'way	−.0005	.0001	−.0006	−.0004
Distance CBD	−.0002	.0005	.00008	−.00005
$LR\chi^2$	41,307			
Pseudo R^2	0.353			
Log likelihood	−33,782			

All continuous variables in ln
Baseline category – urbanized area
*All coefficient significant unless indicated

1. The effect of a 5 % increase in land value in 2009: on the likelihood of land cover change from bare soil to urbanized in 2023
2. The effect of a 10 % increase in residential density in 2009 on land values in 2023
3. The effect of 50 % in decrease in residential density in 2016 on the ratio change in the tendency of land cover changing from fields to urbanized in 2023.

The effects of simulation1 are graphically depicted in Fig. 5. This represents a dynamic version of the basic causal model where land values drive land use which in turns drives land cover. In this instance however we include feedback loops both temporal and spatial in the process. The output relates to the likelihood of bare soil transitioning to an urban (man-made) land cover. As can be seen, much of the change is likely to take place in the south of the metropolitan fringe with the most intensive pockets of change in the east. This reflects the pattern of urban development at the outer edge of the area suggesting creeping urbanization through spillover. Interestingly, public sector initiated urban expansion (such as in Modiin) displays less pressure for creeping development. Alternatively, it could be that initial land values in Modiin in 2009 were lower and thus the pressure for land cover change in the future is less intense.

In simulation 2 we observe the pervasive effects of increasing residential density (Fig. 6). Not only does this have a knock-on effect on land values, by changing land cover and thereby affecting land values in the next time period, but the area affected by this change is also much larger. In comparison with the first simulation where only 8.15 % (50.4 km^2) of total fringe land cover is affected, under this simulation this share reaches 18.3 % (113.2 km^2) Land values increase most in the established urban centers located in the metropolitan fringe. For example, Ashdod and the Arab towns of Tira and Taibe are expected to experience the most serious impacts on future land values.

Finally, simulation 3 deals with the unlikely scenario of residential density decreasing system-wide (Fig. 7). We adopt a particularly extreme scenario that

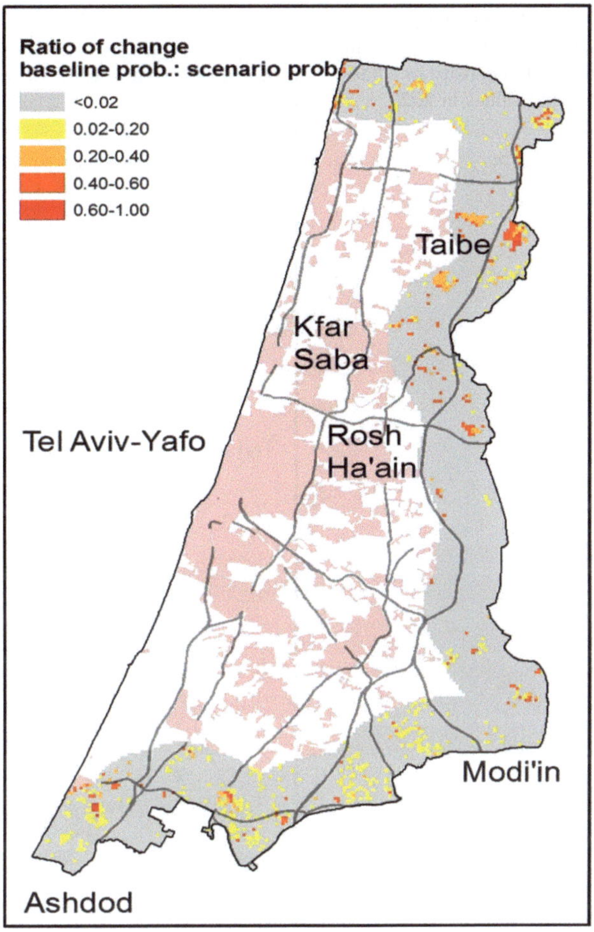

Fig. 5 Simulation 1: the effect of a 5 % increase in land value in 2009 on the land cover tendency to change from bare soil to urbanized in 2023

relates to the effect of a 50 % decrease in residential density on the log odds of fields staying in their current state (or changing to the baseline state of urbanized area) in the next time period. This could only result from some regulatory process such as the imposition of an urban growth boundary or the strict enforcement of green belt rules in the urban fringe that would lower the relative density of development with respect to non urban fringe areas or with respect to non- designated land within the fringe.

Figure 7 shows the ratio change in the tendency of a field cover to become urbanized given the exogenous shock. As can be seen, the ratio changes are as expected mainly negative. Their total area sums to 30.4 km^2. This indicates that drastically curtailing residential development will preserve field land cover. However there are still high level pressure locations where decreasing residential development still does not necessarily reduce all the pressure on field conversion. These include the suburban areas astride the metropolitan fringe such as Shoham,

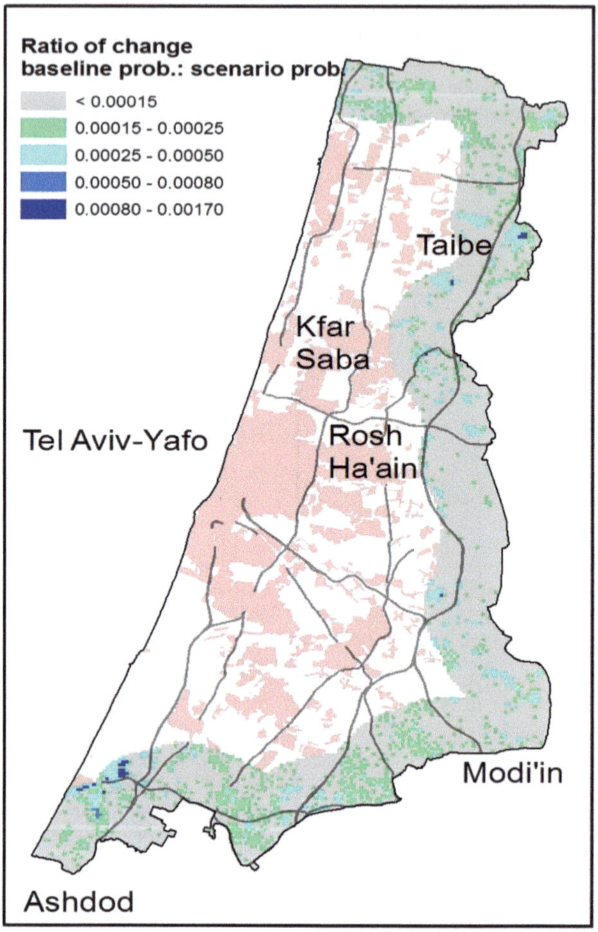

Fig. 6 Simulation 2: the effect of a 10 % increase in residential density in 2009 on land values in 2023

the Modiin area (see Fig. 7 inset) and the areas between the cities of Ashdod and Yavneh. Other affected cells seem to be relatively dispersed and random which seems intuitively correct. The total area of these positive values is 16.7 km^2. Thus, even under a regime of strict development control we still observe points of intense development pressure close to urban centers, in high profile developments along the inner most sections of the metropolitan fringe.

7 Conclusions

This contribution is both methodological and substantive. In terms of method, while short of a full-blown micro-simulation, we have illustrated an approach that combines statistical estimation and impulse response simulation. Akin to economists'

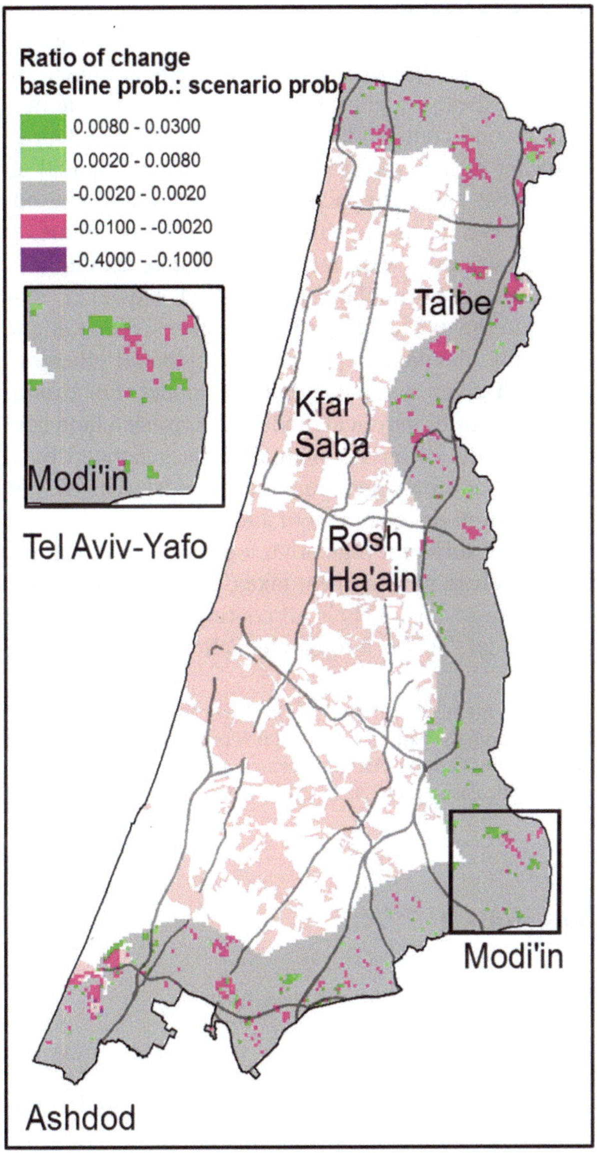

Fig. 7 The effect of a 50 % decrease in residential density in 2016 on the ratio change of land cover tendency to be fields (versus the baseline category of urbanized) in 2023

preferences for explicit representation of land markets, we model land use and land value as occurring simultaneously. Despite our recourse to UrbanSim for generating synthetic data, our simulation is not agent-based we use statistical modeling and spatial simulation to illustrate how temporal and spatial dynamics can be incorporated into the study of land use-land cover interactions. Inevitably, this necessitates addressing issues such as endogeneity and spatial dependence that are inherent in such an exercise. This has involved finding ad-hoc solutions and justifications for some of the shortcomings highlighted above ranging from generating the requisite data, dealing with a very short time series through to successfully identifying the effects of neighbors. Given the methodological motivation underscoring the research, these are areas that will need further attention in the future.

A second contribution has been substantive. Our results point to the precarious character of land cover in the urban fringe. Spatial and temporal feedbacks along with direct causal impacts all work to reduce the odds of contemporaneous land cover surviving in its current state into the future. Admittedly, we have chosen perhaps the most extreme empirical case available in Israel where development pressures are particularly pronounced. However, evidence from other areas also points to the existence of these pressures, albeit at a lower level of intensity.

Finally, the paper highlights the potential inherent in expanding our framework approach given a suitable analytic model. For example, when combined with an ecological or biophysical model the approach here could be extended to forecasting other aspects of environmental stress generated by metropolitan expansion. The urban impacts on vegetation or ecological dynamics or the effect of different forms of residential development of surface hydrology, are obvious examples. The policy impacts of different regulative tools both physical, such as growth boundaries and fiscal, such as development taxes, could also be assessed for their biophysical and ecological results. This is all contingent on integrating with an appropriate evaluative model. This is a research challenge that lies ahead.

Acknowledgements Thanks to Dai Xieer for assistance with the estimation.
This research is partially based on work done in the SECOA (Solutions for Environmental Contrasts in Coastal Areas) research project funded by the European Commission Seventh Framework Programme 2007–2013 under Grant Agreement No 244251.

Appendix 1

Writing Eqs. 5 and 6 in matrix form:

$$U_t = \mathrm{A}_N U_{t-1} + \lambda W U_{t-1} + \delta V_t + K X_t + v_t$$
$$C_t = \Gamma_M C_{t-1} + \alpha U_{t-1} + \lambda U^* + K X_t + u_t$$

where: $\mathbf{A}_\mathrm{N}$, $\boldsymbol{\Gamma}_\mathrm{M}$ are probability transition matrices for LU and LC.

Using the time series lag operator (L) which operates on an element, in order to produce the previous element, we solve for $\mathrm{U_t}$ and $\mathrm{U_{t-1}}$:

$$(I - \boldsymbol{\Gamma} L) C_t = B U_{t-1} + u_t \quad (9)$$
$$(I - \boldsymbol{A} L) U_t = v_t \quad (10)$$

$$U_t = (I - AL)^{-1} v_t \tag{11}$$

$$\begin{aligned}(I - AL)^{-1} &= I + AL + A^2L^2 + \ldots \\ A^n &\to 0 \; n \to \infty\end{aligned} \tag{12}$$

$$U_{t-1} = v_{t-1} A v_{t-2} + A^2 v_{t-3} + \ldots \tag{13}$$

Substituting for U_{t-1} in Eq. 9 and re-arranging:

$$C_t = (I - \Gamma L)^{-1} B\left(v_{t-1} A v_{t-2} + A^2 v_{t-3} + \ldots . \right)$$

The inverse of the polynomial is:

$$(I - \Gamma L)^{-1} = I + \Gamma L + \Gamma^2 L^2 + \Gamma^3 L^3 + \ldots$$

Appendix 2: Probability Estimation of Land Cover Data

We use MNL regression for historic (2002) data to create log odds of land cover (LC) e change (from bare soil, forests, orchards, fields to built areas). Independent variables are: LC for 1995 (lc95), the most frequent LC in the neighboring cells in 19955 (neigh95), distance from 2 digit highways (2dgt), distance from 1 digit highways (1dgt), distance from metropolitan CBD (core), parks area (prk), 1995 residential land value (lv95), 1995 commercial square feet (comm95), 1995 industrial square feet (ind95), 1995 governmental square feet (gov95), 1995 number of residential units (res95). Non –significant variables are excluded and the regression re-estimated with 0.2445 misclassification rate which is equal to 51.1 % accuracy. Coefficient values are as follows:

Log odds for	Intercept	2dgt	1dgt	core	prk	Lc95[2]
2/6	1.06300627	0.02128593	−0.0212716	−0.0000239	−0.000018435	1.39784825
3/6	1.6741873	0.0547843	−0.0546622	−0.000070779	−0.0001232	0.00525623
4/6	0.01532415	0.01373619	−0.0138219	−0.0000009686	−0.0001346	−0.8704572
5/6	0.11464507	0.02095321	−0.021061	−0.0000082171	−0.000061941	−0.0847443

Log odds for	Lc95[3]	Lc95[4]	Lc95[5]	neigh95[2]	neigh95[3]
2/6	1.3162059	−0.5565036	−0.9872424	0.81029937	0.24571899
3/6	2.21462143	0.91567457	−0.8596639	−0.1341568	1.35462331
4/6	−0.6881597	3.05639013	−0.1230151	−0.2588099	−1.0215386
5/6	−1.5210677	0.23063331	2.64963015	−0.0652665	−1.1603799

Log odds for	neigh95[4]	neigh95[5]	lv95	res95
2/6	−0.0369579	−0.4688555	−0.0000000056334	−0.0053964
3/6	0.33465383	−0.9959705	−0.0000038221	−0.0217897
4/6	1.39302637	0.11419243	−0.0000018486	−0.0196984
5/6	0.19326558	1.08702141	−0.0000033074	−0.016155

These parameters are then tested by using the 2002 LC data and 2002 land use model outputs as independent variables. The predictions of this regression generate 53.48 % accuracy when compared with actual 2009 LC data. After each round, the outputs of the previous round become the *t-1* values for a new round, i.e. LC value calculated in round t becomes LC data for round $t + 1$. This is repeated for each time period with time lags of t = 7.

References

Alberti, M. (2005). The effects of urban patterns on ecosystem function. *International Regional Science Review, 28*(2), 168–192.

Alberti, M. (2008). *Advances in urban ecology: Integrating human and ecological processes in urban ecosystems*. New York: Springer.

Alberti, M., & Waddell, P. (2000). An integrated urban development and ecological simulation model. *Integrated Assessment, 1*, 215–227.

Baltagi, B. H. (2008). *Econometric analysis of panel data*. London: Wiley.

Felsenstein, D., & Ashbel, E. (2010). Simultaneous modeling of developer behavior and land prices in UrbanSim. *Journal of Transportation and Land Use, 3*(2), 107–127.

Felsenstein, D., Ashbel, E., & Ben-Nun, A. (2007). Microsimulation of metropolitan employment deconcentration. In E. Koomen, J. Stillwell, A. Bakema, & H. J. Scholten (Eds.), *Modelling land use change: Progress and applications, volume 90 of GeoJournal library* (pp. 199–218). Netherlands: Springer.

Haan, P., & Uhlendorff, A. (2006). Estimation of multinomial logit models with unobserved heterogeneity using maximum simulated likelihood. *The STATA Journal, 6*(2), 229–245.

Hepinstall, J. A., Alberti, M., & Marzluff, J. M. (2008). Predicting land cover change and avian community responses in rapidly urbanizing environments. *Landscape Ecology, 23*, 1257–1276.

Ihlanfeldt, K. R. (2007). The effect of land use regulation on housing and land prices. *Journal of Urban Economics, 61*, 420–435.

Irwin, E. G. (2010). New direction for urban economic models of land use change: Incorporating spatial dynamics and heterogeneity. *Journal of Regional Science, 50*(1), 65–91.

Irwin, E., Jayaprakash, C., & Munroe, D. K. (2009). Towards a comprehensive framework for modeling urban spatial dynamics. *Landscape Ecology, 24*, 1223–1236.

Lee, M. J. (2013). Practical panel conditional logit estimators for binary, Ordered discrete and multinomial responses. *The Oxford handbook of panel data econometrics*. Oxford University Press, Oxford, UK.

Levin, N., Harari-Kremer R., & Carmel, Y. (2011). *Remote sensing of Israel's natural habitats*, Unpublished Report for the Israeli Nature and Parks Authority, Jerusalem.

Lopez, A., Bocco, G., Mendoza, M., & Duhau, E. (2001). Predicting land-cover and land-use change in the urban fringe: A case in Morelia city, Mexico. *Landscape and Urban Planning, 55*, 271–285.

Mayer, C. J., & Somerville, C. T. (2000). Land use regulation and new construction. *Regional Science and Urban Economics, 30*, 639–662.

Neyman, J., & Scott, E. L. (1948). Consistent estimates based on partially consistent observations. *Econometrica, 16*, 1–32.

Parker, D. C., Manson, S. M., Janssen, M. A., Hoffman, M., & Deadman, P. (2003). Multi-agent systems for the simulation of land-use and land-cover change: A review. *Annals of the Association of American Geographers, 93*, 314–337.

Ravetz, J., Fertner, C., & Nielsen, T. (2013). The dynamics of peri-urbanization. In K. Nilsson, S. Pauleit, S. Bell, C. Aalbers, & T. Nielsen (Eds.), *Peri-urban futures: Scenarios and models of land use change in Europe* (pp. 13–44). Heidelberg: Springer.

van Schrojenstein Lantman, J., Verburg, P., Bregt, A., & Geertman, S. (2011). Core principles and concepts in land use modelling: A literature review. In E. Koomen & J. Borsboom-van Beurden (Eds.), *Land use modelling in planning practice* (pp. 35–57). Dordrecht: Springer.

Verburg, P. (2006). Simulating feedbacks in land use and land cover change models. *Landscape Ecology, 21*, 1171–1183.

Verburg, P. H., Schot, P. P., Dijst, M. J., & Veldkamp, A. (2004). Land use change modeling: Current practice and research priorities. *GeoJournal, 61*, 309–324.

Waddell, P., & Ulfarson, G. F. (2003). Introduction to urban simulation: Design and development of operational models. In P. Stopher, K. Button, K. Kingsley, & D. Hensher (Eds.), *Handbook in transport, vol 5. Transport geography and spatial systems*. New York: Pergamon Press.

Lichens and Plants in Urban Environment

Gregory E. Insarov and Irina D. Insarova

Abstract This chapter considers alteration in lichens and plants caused by urbanization, an increasingly prominent driving force shaping the Earth's landscape. The chapter does not presume to cover all the aspects of plants and lichens change in context of urbanization. Its main focus is a short general overview of the main processes taking place in urban lichens and plants on level of organism, community and ecosystem, and of ecological services they provide. Plant species richness and composition, community and ecological processes, growth, anatomical, morphological traits and physiological processes under urbanization stress are discussed. Ecological services of city vegetation: air quality improvement, greenhouse effect reduction, abatement of urban noise, among other, are described. Methodological aspects of air pollution impact assessments with lichens are discussed. A number of case studies fulfilled in cities across the world to assess change of urban lichen communities are reported. The importance of lichens as monitors of temporal and spatial trends in urban biota state under air pollution stress is justified. City planning and management implications are provided.

1 Introduction

Global urban population increased from 2 % in the beginning of the twentieth century to 50 % in the beginning of the twenty-first century (UNDESA 2011; Lynch 2012). Percentage of urban dwellers even higher in some countries, for instance in the UK, France and USA it is about 90 %. The level of global urbanization is

G.E. Insarov (✉)
Institute of Geography of the Russian Academy of Sciences, 20-b Glebovskaya Street, Moscow 107258, Russia
e-mail: insarov@lichenfield.com

I.D. Insarova
Department of Biology, Moscow State University, Vorobievy Gory, Moscow 119899, Russia
e-mail: insarovaid@mail.ru

D. Malkinson et al. (eds.), *Modeling of Land-Use and Ecological Dynamics*, Cities and Nature, DOI 10.1007/978-3-642-40199-2_9, © Springer-Verlag Berlin Heidelberg 2013

expected to rise from 52 % in 2011 to 67 % in 2050 (UNDESA 2011). The human habitation of the Earth has shifted from predominantly rural to majority urban over the last century (Pickett et al. 2011).

Urban ecosystems are those in which people live at high densities, and where built structures and infrastructure cover much of the land surface. Urbanization can be defined as the process of installation of anthropogenic structures (e.g., buildings and roads) in existing natural or farming areas, in order to satisfy human population requirements (Croci et al. 2008; Pickett et al. 2011). According to this definition, the urbanization is an increasingly prominent driving force shaping the Earth's landscape. Urbanization causes deep modification of the land use and land cover, affects climate conditions, energy flows, biochemical cycles, soil properties and hydrology.

The fringes of cities and rural regions within commuting distance of cities, demonstrate some of the highest rates of development in the world (Czamanski et al. 2008). As a result, significant changes to the landscape and the alteration of ecosystems functions and biodiversity appeared. As suburbia spread, opportunities come for some species to exploit new resources, and investigations of differences in the composition of communities along urban to rural gradients have produced contradictory results: while some communities and species demonstrate clear decline from rural to urban end of the gradient, others show high abundance in the intermediate suburban area followed by decreasing abundance both in rural and urban areas (DeStefano and DeGraaf 2003; Rango 2005; Blair 1999). Thus, the effects of urbanization on community composition may not be easily and deterministically projected.

Despite the prevalence and acceleration of urbanization, its effects on biological communities are still not clear. Urbanization is accompanied by habitat fragmentation, air, soil and water pollution, anthropogenic climate change (heat island effect, e.g. Oke (1995), warmer winters and, as a result, a longer vegetation period), species introduction, noise. These factors affect communities in urban environments and in combination with ground and air temperature, relative humidity, edaphic conditions, amount of prey/predator items, to name some, all modify the natural ecosystems in the urban environment.

Direct effects of urbanization are masked by intrinsic species variation. To detect the effect of a given factor, e.g. of air pollution on urban biota, we should select a group of organisms which combine sensitivity to this factor with relatively high intrinsic stability. For air pollution, such a group is lichens (Hawksworth 1971; Nimis et al. 2002; Purvis 2007; Insarov et al. 2010). Combination of sensitivity to air pollution and intrinsic stability is important for detection of signal (response to air pollution) masked by noise (variability due to factors other than air pollution). Besides, we consider plants, including trees because of their important role for people in cities and for urban ecosystems as well.

Urban ecology traditionally studies distribution and abundance of organisms in and around cities. Important application of urban ecology is urban planning where urban ecology results are used on designing the environmental amenities of cities for people, and on reducing environmental impacts of urban regions (Pickett et al. 2011). Implications of urban ecology science for urban planning are considered in this chapter.

The purpose of this chapter is (1) to discuss alteration of plants under urban stress, and (2) to synthesize studies on lichen communities alteration in space and time in context of urban air quality change. Implications for urban planning and management are also discussed.

2 Plants in Urban Environment

There are many reasons why plants are important for urban dwellers. Atmosphere in cities is enriched by oxygen from one hand, and carbon dioxide level is reduced from other hand because of plant photosynthesis. Ozone level also declines as tree-covered area increases, this was demonstrated for the US cities of the eastern seaboard from Washington, DC, to central Massachusetts (Nowak et al. 2000; Luley and Bond 2002). Reduction of carbon dioxide and ozone level mitigates greenhouse effect. Plants improve urban air quality because gaseous pollutants, such as SO_2, NO_x, CO, are absorbed by vegetation, and particular matters (dust, heavy metals) are deposited on leaves and immobilized.

In cities, plants mitigate gusting wind and shade buildings preventing them from superheating in summer. Vegetation improves summer urban climate providing cooling in result of solar radiation absorption by tree canopies and of evapotranspiration from leaves (Nowak and Dwyer 2007; Kato et al. 2013). Owing to vegetation, peak summer temperature can be reduced for 1–5 °C, depending on land-use type (US EPA 2013). Trees hamper heavy rainfalls, thereby decrease loss and damage caused by these extreme events. Storm water intercepted by tree crown and stem, drops down or evaporates. Amount of intercepted water depends on tree species, tree height, stem and leaves area, foliation period, crown density (McPherson et al. 2001).

Noise is also abated by trees. Urban noise can reach unhealthy level and exceed 100 dB. Vegetation barrier in conjunction with relief can reduce noise by 6–15 dB. Plants absorb high frequency noise which is most distressing to people (McPherson et al. 2001).

Urban parks and forests are the main habitat for animals, including birds, mammals and arthropods. Green areas improve physical and mental health of human being, provide aesthetic surroundings and reducing stress of urban life. Parks and gardens are important recreational zones for urban population (Nowak and Dwyer 2007; Tzoulas et al. 2007).

2.1 Species Diversity and Plant Growth Along Urban–Suburban Gradient, Native Versus Alien Species

Plant biodiversity is influenced by urbanization because urbanization is accompanied by temperature and precipitation increase, air, water and soil properties change, habitat fragmentation and destruction. Habitat fragmentation disrupts seed dispersal and reduces range for dispersal-limited plant species (Kühn and Klotz 2006; Kowarik et al. 2011).

Species diversity in cities of Europe and North and South Americas is higher than in surrounding areas (e.g., McKinney 2002; Hope et al. 2003; Kühn et al. 2004; Knapp et al. 2009; Kühn and Klotz 2006; Leveau and Leveau 2005) because of considerable fraction of alien and introduced plant species. Increase of alien plant species and declines of native species increase genetic, taxonomic and functional similarity between different urban regions (Kühn and Klotz 2006). This process is known as biotic homogenization (Olden 2008).

So, plant species diversity is higher in urban areas comparing with rural ones. Biodiversity, however, includes also diversity of phylogenetic relationships in plant communities. For Germany, it was demonstrated that plant phylogenetic diversity in cities is somewhat less than that in rural areas, in opposite to species diversity. This means that plant species in cities are more close relatives than outside cities. Because their genetic pool is more homogeneous, plant community ability to response to environment change is reduced. Loss of phylogenetic information may have negative influence on ecosystem functioning (Knapp et al. 2008).

Average fraction of alien plant species in total plant species in 54 small cities in Central Europe (25 Polish, 24 German, 4 Czech and 1 Austrian), was 40 % (Pyšek 1998). In Central England cities, top 20 urban plant species ranked by occurrence are aliens, whereas from top 20 plant species inhabited urban-and-vicinity areas, 18 are native to Britain (Hill et al. 2002).

It was demonstrated with the example of German cities that urbanization causes homogenization of both total plant species flora and native plant species. This is because alien species invasion into cities is accompanied by extirpation of native plant species, in particular of rare ones. Protection of urban plant habitats is needed to prevent flora homogenization (Kühn and Klotz 2006).

Change of vegetation quality in Mediterranean urban region in context of cities growth was demonstrated with example of Rome. The city experienced intensive growth in 1960–2006 at the expense of adjoining forests, pastures and wetlands. As many other Mediterranean cities, modern Rome development can be divided into two phases: 'compact growth' which took place until 1990s, and low-density expansion, or 'sprawl'. Built-up area increased from 3.3 % of present area in 1960 to 12.9 % in 2006. Vegetation quality (VQ) has been estimated by a number of indices, including vegetation resistance to drought and vegetation cover. VQ decreased along rural–urban gradient. Both in the rural and in the urban areas VQ was worsened during the 'compact growth' phase and slightly improved during the 'sprawl' phase. During both phases, vegetation quality was lower in suburban than in rural areas (Salvati and Zitti 2012).

Urbanization-driven change of forest and soil was studied along transect from New York City to rural Litchfield County, CT (McDonnell et al. 1997). Urban red oak forests have lower stem density, impoverished understory and more non-native sapling species than similar stands in the rural areas. Rate of litter decomposition and nitrification was higher in urban forests than in rural ones, and soil properties were different: in urban soils labile carbon pool was smaller than in rural ones, but urban soils' total passive soil carbon pool was greater than rural ones. This is attributed to lower activity of forest soil microorganisms in urban areas.

High air pollutant content, especially of sulphur dioxide and dust in big cities in the mid twentieth century caused growth impairment and death of sensitive herbaceous and woody plants (e.g. Kunina et al. 1979; Guderian 1977; Bell et al. 2011). Level of SO_2 and of some other airborne pollutants drastically decreased in European and North American cities by 1990s, and urban trees in developed countries have higher growth rate than non-urban ones (Nowak and Crane 2002).

Concentrations of air pollutants in cities are higher than in rural areas with one important exception, namely concentrations of ozone (O_3) are usually higher in suburban and rural areas. Ozone is a product of photochemical reactions with substrates including volatile organic compounds (VOCs) and nitrogen oxides (NO_x). Ozone level in cities usually is lower than in city vicinities because it is destroyed by NO emitted by motor vehicles (Bell et al. 2011), e.g. in three Italian cities, Milan, Florence and Bari it was shown that average O_3 level in cities was 2.6 times lower than in rural and suburban areas (Paoletti 2009). Biomass of cloned cottonwood saplings (*Populus deltoides* Marsh.) at the end of vegetation period in surrounding rural areas of New York City was lower than in NYC itself. This is attributed to the higher O_3 concentration in ambient air of rural areas (Gregg et al. 2006). Searle et al. (2012) studied red oak (*Querqus rubra* L.) seedlings along urban – suburban – rural transect also originated in NYC. Oak biomass in NYC Central Park was eight times higher than in rural sites. Urban plants had greater shoot/root ratio than rural ones, consequently photosynthetic total leaf area of urban plants also was greater. Ozone levels are not considered as elicited a response in red oak growth. Another factor changing along the gradient and influencing tree growth is air temperature. Importance of higher temperature in NYC, especially high night-time temperatures, comparing with suburban and rural sites (difference in temperature between urban and rural sites was 2.4 °C for average maximum temperatures, and 4.6 °C for average minimum temperatures) is emphasized (Searle et al. 2012).

Tree growth inter-comparison for small, medium, large towns and rural areas was conducted in five U.S. midwestern states (Iakovoglou et al. 2002) for silver maple (*Acer saccharinum* L.), honeylocust (*Gleditsia triacanthos* L.), hackberry (*Celtis occidentalis* L.), black maple (*Acer nigrum* Michx. F.), and basswood (*Tilia americana* L.). Mean growth rate in cities of all types was higher than in rural areas, while no differences were revealed between cities of all types. Growth rate of young trees was higher in all habitats than that of older trees. City parks provided higher tree growth rates than downtown sites, growth rate in residential areas was intermediate.

Plant growth depends on both abiotic factors (such as climate, air quality, soil physical and chemical properties, including pH and bulk density, availability of nutrients, especially nitrogen and phosphorus) and biotic ones (position of plant in community: competition for light and space for roots, soil macrofauna and microorganisms, insects, arboreal fungi, pathogens, etc.). In cities, salinity caused by de-icing salts influence tree growth as well (Fostad and Pedersen 2000).

2.2 Combined Effect of Air Pollution and Pathogens on Plants

Both natural and urban plants are subjects to invertebrate pest attacks and infection by microorganisms (fungi, bacteria and viruses), but for all that insect pests are more active in cities, e.g. review by Flückiger et al. (2002). Change of air chemistry caused by air pollution likely contributes to higher sensitivity of urban plants comparing with natural ones.

Fungal pathogens are also more active in polluted environment of cities, however plant – fungal pathogens interaction is more ambiguous (Beckett et al. 1998). In Newcastle region, UK the fungus *Rhytisma acerinum* (Pers.) Fr., an agent of the leaf disease of sycamore (*Acer pseudoplatanus* L.) known as tar spot, was studied by Bevan and Greenhalgh (1976). The pathogen is rather sensitive to SO_2 level in the air, it has not been recorded in areas stressed by SO_2 in concentration higher than 90 μg/m^3. Decline of sulphur dioxide level in cities can be a reason of outbreaks of the sycamore leaf disease. Nowadays, a strong inhibition effect of nitrogen oxides on *R. acerinum* growth and development is revealed in London. Level of host tree infection by *R. acerinum* increased along transect from London center to vicinity, this can be explained by high concentrations of nitrogen compounds in the urban air (Bell et al. 2011).

In UK, laboratory and field studies of the fungus *Diplocarpon rosae* Wolf producing the blackspot disease of roses demonstrated that fumigation by SO_2 having concentration 100 μg/m^3 in course of 2–10 days, depending on the region, causes almost 100 % inhibition of infection (Saunders 1966). Blackspot disease of roses in UK cities in 1960th has not took place, however there are many records of the disease nowadays, in line with decrease of sulphur dioxide level (Bell et al. 2011).

2.3 Anatomical, Morphological and Physiological Response of Plants to Air Pollution

Air pollution impacts on plants directly and indirectly. Initial targets of direct impacts are processes in plant cell, e.g. photosynthesis, and cytological features as well. Direct impacts of air pollution can be divided on chronic and acute ones. Acute effects result from exposure to a high concentration of pollutant over a relatively short time and are manifested in the form of visible symptoms, usually in the form of leaf and needle lesions. In contrast, chronic damage results from prolonged exposure to lower concentrations and is manifested as reduced growth and/or yield, often without visible symptoms. Indirect impacts include altered sensitivity to other stresses and changes in soil chemistry and microbiology. Indirect impacts include also changes in plant sensitivity to abiotic factors and biotic ones in the form of pest attack or pathogen infestation. Indirect impacts of air pollution can be via soil (acidification, eutrophication) and via changes in diversity and number of soil microorganisms, including mycorrhizal fungi.

Plants are efficient in trapping and accumulation of metals, dust, organic compounds etc. The elements are accumulated in different organs and tissues of plants with various consequences for the organisms. The bioaccumulation of elements from air, water and soil are widely studied nowadays in cities. Tree foliage, in particular, is regarded as a good monitor of the environment.

Morphological and anatomical features of plants change along with ambient conditions, there are used as monitors of urbanization. In Gent, Belgium, leaf features of herbaceous plants dandelion *Taraxacum officinale* Wigg. and plantain *Plantago lanceolata* L. were studied in urban (harbor, roadsides, industrial zones) and suburban (natural, agricultural, urban green) areas (Balasooriya et al. 2009; Kardel et al. 2010). It was demonstrated that the mean stomatal pore surface decreased from suburban to urban areas, whereas stomatal density increased both on the adaxial and the abaxial leaf surface. Calculated minimal stomatal resistance increased from suburban to urban areas, i.e. decreased stomatal size outweighs increased stomatal density. Mapping of stomatal characteristics was elaborated. Both plants demonstrated clear morphological and anatomical response to differences in habitat quality, they can be used as urban air quality monitors. Stomata play key role in plant gas exchange with the environment, and increased stomatal resistance in urban areas is a mechanism to control gas exchange and to reduce uptake of air pollutants (Larcher 2003).

The same pattern of stomatal traits was found in comparative study of white willow (*Salix alba* L.) in two other cities in Belgium, Antwerp and Zoersel. The cities are different in terms of urbanization: Antwerp has a high population density and industrial activity, while Zoersel has no industry, moderately populated and has vast forest land. Stomatal length and width in Antwerp city were certainly lower than in Zoersel, and stomatal pore surface 20 % less than in Zoersel, while stomatal resistance in Antwerp city was 17 % greater. In spite of the fact that shoot biomass and specific leaf area (ratio of total single sided fresh leaf surface area to its oven dry weight) were similar in both cities, it is demonstrated that white willow can adapt to urban environment generating many small stomata what causes increase in stomatal resistance (Wuytack et al. 2010).

Cuticle and epicuticular wax layer play important role for stomata functioning, prevent water loss, protect leaves from infections by pathogens. Air pollutants change structure and chemistry of epicuticular wax layer (Kardel et al. 2012). Epicuticular wax layer determines leaves wettability, it is species-specific and depends on epidermis features and leaf shape. In Gent, leaves wettability change under urbanization stress was studied for five wood species, *Alnus glutinosa* (L.) Gaertn., *Acer pseudoplatanus* L., *Betula pendula* Roth, *Quercus robur* L. and *Sambucus nigra* L. Leaves shape and size as well as leaves surface properties are different for these five species. Measure of leaf wettability is the angle between perimeter of drop on the leaf surface and the surface itself, this angle is named drop contact angle (DCA). The greater is DCA value, the smaller is wettability. Effect of urbanization on leaves wettability of *B. pendula* and *A. pseudoplatanus* was masked by a strong factor acting in all habitats: thin flexible twigs of birch are rubbing each other by wind, this causes leaf-to-leaf contacts and abrasion of the epicuticular wax. Leaf wettability of *A. glutinosa* significantly depended on sampling time (June vs. September), whereas wettability difference in different habitat types was insignificant. Significant change of wettability in response to habitat quality change was revealed for two of five wood species, namely for *Q. robur* and *S. nigra* (Kardel et al. 2012). For *Q. robur* leaves, DCA value decreased from semi-natural areas to industrial zones, this is possibly because of epicuticular wax layer damage caused by stress. On the contrary, DCA value for *S. nigra* leaves decreased in opposite direction. This can be explained by influence of other factors, namely by shading, because *S. nigra* is a shrub species much lower than *Q. robur*, it grows under the tree shade in semi-natural areas while it is well exposed in urban areas (Kardel et al. 2012), and leaf wettability certainly increases in shading habitats comparing with sun-exposed ones (Pandey and Nagar 2002).

In Helsinki Metropolitan Area, southern Finland, leaf traits of European aspen (*Populus tremula* L.) in three urban and three rural forest stands were compared (Nikula et al. 2010). Specific leaf area was about the same, however urban leaves contained more epicuticular waxes. Nitrogen content was higher in urban leaves, in line with higher N deposition, carbon content was similar. Urban leaves litter contained higher level of nitrogen and lower level of lignine and phenols, this make it relatively easy cleavable and attractive for decomposers. The main adaptation to urban conditions of *P. tremula* in Helsinki is increase of epicuticular waxes in leaves (Nikula et al. 2010).

2.4 *Role of Plants in Mitigation of Urban Air Pollution and Carbon Storage and Sequestration*

Dense and branched road network in modern cities and traffic influence urban biota. Roadside vegetation is impacted by mechanical damage, de-acing agents, dust, gaseous pollutants, heavy metals. Besides, roads contribute to area fragmentation. From other hand, plant migration is facilitated by roads because they provide suitable migration corridors for seeds and propagules transferred by vehicles. Alien plant species are very common along roadsides (e.g. von der Lippe and Kowarik 2007).

Dust particles can be soil derived, other are produced by mobile and stationary emission sources. Particles accumulation by oak leaves on different distance from the highway was studied in Rough Wood, Walsall, UK (Freer-Smith et al. 1997). It was found that a large fraction of particles were organic, majority of inorganic particles were soil derived while others were products of combustion or local metal works. The number of particles on leaves was higher for trees near the road than for more distant trees. Dust, particularly from roads, is a pollutant which is injurious to human health and blights plants. Roadside vegetation improves air quality by air filtration and immobilization of the dust particles.

Gaseous pollutants and dust are accumulated in plants and cause morphological, anatomical, biochemical and physiological changes (Kunina et al. 1979; Beckett et al. 1998; Nowak and Dwyer 2007). Pollutant accumulation power varies across plant species and depends on climatic, soil and other features of habitats. Gaseous pollutants are removed by vegetation mainly by uptake via stomata, whereas airborne particles are intercepted by trees, shrubs and grass. Particles are washed out from plants by rain, or fall to the earth with leaves and twigs.

Annual removal of carbon monoxide (CO), nitrogen dioxide (NO_2), ozone, particulate matter less than 10 mm (PM_{10}) and sulfur dioxide (SO_2) were estimated for 55 cities and for other urban areas across the lower 48 United States based on canopy architecture and pollutant deposition modeling. Annual pollution removal varied greatly among the cities, total removal estimate for all urban trees in the coterminous United States in 1994 was 711,000 t. Pollution removal was monetized using externality values for the United States for each pollutant, total annual value estimate was $3.8 billions (Nowak et al. 2006). This value is significant enough to consider management of city green areas and roadside vegetation as important tool to improve nationwide urban air quality.

In Guangzhou city in South China, value of pollution removal by trees was also estimated. Rapid urbanization, industrialization and migration from rural to urban areas in China in last decades resulted in increase of airborne pollutants concentrations and overall air quality deterioration (Jim and Chen 2008). Main emission sources in Guangzhou are industry, motor transport and residential areas, primary pollutants are SO_2, NO_x and total suspended particulates. In according to 2000 inventory, it was 1,794,455 urban trees in Guangzhou. Estimate of total annual removal by trees of main pollutants in Guangzhou city 312.03 t in 2000. Pollutant concentration, proximity of trees to emission sources and tree covered areas are crucial for efficient pollutant removal by trees. To estimate the monetary value of this ecosystem service, marginal costs were used. The total annual value of air pollutant removal by urban trees in Guangzhou city was estimated at $11,000 in 2000. This estimate is much less than value of air pollutant removal by trees estimated for US cities, for instance for Philadelphia and New York city in 1994 these estimates were $5.13 millions and $9.24 millions, correspondingly (Nowak et al. 2006). The difference can be explained by much lower marginal cost for air pollutant removal in China than in US, usage of advanced (and more expensive) technologies for air pollution control in the US, as well as lower labor and material costs in China (Jim and Chen 2008). Monetary estimate of important ecosystem service can provide a basis for justification of investments to the urban greenery as an important part of urban infrastructure in China and in rapidly growing cities in general.

Global concentration of carbon dioxide, the most important anthropogenic greenhouse gas has been increasing: it was about 280 ppm in pre-industrial period (before 1750), and 379 ppm in 2005 (Solomon et al. 2007). The main source of this increase is fossil fuel use, land-use change, including deforestation also makes significant contribution. Greenhouse gas concentration increase is likely to be the main cause of global warming. Trees are CO_2 sinks because they fix carbon in course of photosynthesis and store excess carbon as biomass. Long-term dynamic of carbon storage and emission by forest change because trees grow, die and decay.

Estimate of carbon amount stored by urban trees in coterminous USA cities was obtained on the base of field data from ten US cities and on national urban tree cover data (Nowak and Crane 2002). The total estimate was 700 million tons of carbon. Carbon storage varied between cities: from 1.2 million tons in New York and Atlanta to 19,300 t in Jersey City. Gross sequestration rate varied from 42,100 tC/year in Atlanta to 800 tC/year in Jersey City. The greater tree cover and the fraction of large and healthy trees in urban trees population, the higher sequestration rate. Because tree cover in urban forests is lower than in forest stands, carbon storage per hectare by urban forest (typically 25.1 tC) is lower than that by forest stands (53.5 tC). However, carbon storage per unit tree cover by urban trees (9.25 kgC/m^2 cover) is much greater than in forest stands (0.3 kgC/m^2 cover) because of large fraction of large trees and of more open structure of urban forests. The total monetary value of carbon storage by USA urban trees is $14,300 million, with an annual sequestration value of $460 million (Nowak and Crane 2002).

3 Lichens as Monitors of Air Quality in Cities

Lichens are symbiotic organisms comprising association of two or more components, fungi and green algae, or cyanobacteriae. Systematically lichens are included into fungus kingdom, which is separate from plant kingdom. Algae produce organic carbon which fungi consume. Fungus mass is up to 90 % of the lichen weight.

Fungal and algal components should be in equilibrium to provide development of lichen as entire organism. Lichens occur in most terrestrial ecosystems of the world. They are adapted to wide range of environmental conditions varying from hot deserts to polar regions and high mountains harsh conditions. Lichens colonize natural substrates: rocks, soil, tree bark and branches, dead wood, etc., and man-made substrates like concrete and steel constructions. Lichen species and communities vary between regions and substrates.

3.1 Why Lichens Are Chosen Organisms for Air Pollution Studies?

In spite of high adaptability to extreme habitats, lichens are rather sensitive to such environment changes as climate change, habitat fragmentation (Esseen and Renhorn 1998) and gaseous composition of atmosphere alteration, or air pollution. The main reasons for lichen sensitivity to air pollution are as follows:

1. Sources of water and solute substances for lichens are atmosphere, precipitations, fog and dew,
2. Contrary to plants, lichens lack both stoma and protective cuticle, water is absorbed by the whole of outer thallome surface,
3. Lichens accumulate substances they receive from the ambient environment, and concentrations of the substances within the thalli can be rather high,
4. Lichens are perennial organisms and they can metabolize under lower temperature than most of the plants do, i.e. they are damaged during longer period than most of the plants,
5. Equilibrium of fungal and algal components can be easily destroyed by external stress.

Lichen communities combine, in common, high species sensitivity to air pollution with low intrinsic variability. Under given conditions, the cover, frequency and species composition of lichen community are stable due to:

1. Relatively independence from other organisms (they are seldom involved into food webs),
2. Low reproductive rate,
3. Low growth rate,
4. Longevity.

This combination of sensitivity to air pollution and of intrinsic stability is important for detection of signal (response to air pollution) masked by noise (variability under other than air pollution factors). Lichens have low signal-to-noise ratio, and this provide good base for trend detection by lichens comparing with other organisms (Izrael et al. 1985). Lichen community species composition and abundance depend on climate, microclimate, substrate (Barkman 1958) and on concentration of such substances in the atmosphere surface layer as sulphur dioxide, nitrogen compounds, fluorides etc. If concentrations and/or gaseous composition change, lichen communities change as well. These changes can help in determination of environment change direction.

3.2 Urban Lichens and Air Pollution: Research Methodology Development

Poor abundance of urban lichens comparing with suburban and peri-urban ones is known for about two centuries (Brightman 1982; Laundon and Waterfield 2007; Purvis 2010). In 1859, Grindon attributed the decline of some species in the Manchester area largely to the air pollution. Similar finding was made in Paris by Nylander in 1866. Afterwards, numerous studies linked impoverished lichen diversity in cities with air pollution. Urban dry air was considered as another reason for this impoverishment (Hawksworth 1971). Since 1912, Sernander published results of lichen studies in Sweden, he suggested sub-division of areas under air pollution stress into three zones: lichen desert (lavoken), struggle zone (kampzon) and normal zone (normalzon). This division was based primarily on epiphytic lichens: there were no lichens in the lichen desert, crustose lichens, foliose lichens of low vitality and no fruticose lichens can be found in the struggle zone, in normal zone, mainly in parks, fertile lichens with no visible damage of all three growth forms can be found.

Development of zonal system continued in many cities in Europe. Atmospheric pollution alters lichen communities and usually results in an impoverishment in terms of richness and/or abundance. Authors of most studies on lichens and air pollution consider air pollution as the main reason of lichen vegetation change in cities, some authors used lichens as air pollution (AP) indicators for monitoring AP effects, and even for level of AP itself in cities of different size. However, it was recognized that other than AP factors influencing lichen species composition, abundance and vitality should be also taken into account (Hawksworth 1971).

Urban lichen communities were studied intensively in Europe, North America and beyond from 1970th up to nowadays. International programs on lichen monitoring in context of air pollution in Europe, USA, Canada and East Asia were elaborated (e.g. Trunk Epiphytes 2004; Insarov 2007) on the base of agreed methodology of field study and data handling/processing (Nimis et al. 2002 and others). Hawksworth and Rose (1970) paper stimulated high interest to lichen zonation in cities. They demonstrated that there is correlation between mean sulphur dioxide level during periods of greatest humidity (i.e. winter) with lichen distribution. Short term very high sulphur dioxide levels may have relatively little effect. At that time, sulphur dioxide was the main pollutant in cities and around industry enterprises, so city zonation and mapping was based on correlation between mean sulphur dioxide levels and lichen distribution, and air pollution levels for different parts of the city could be compared. Scales of lichen species sensitivity and poleotolerance (the concept of sensitivity is opposite to that of poleotolerance) were elaborated, they mainly are based on lichen appearance/abundance in zones with different air pollution levels (Trass 1973; Insarova et al. 1992; Nimis et al. 2002 and others). It was also recognized that some species appear to have different tolerances in different geographical regions, and a scale drawn up in one region can not consequently be reliably applied to another without prior study (Hawksworth and Rose 1970). Scales of tolerance should be either worked out in the area to be surveyed, or interpolation procedures for hierarchical data should be applied (Roitman 1989).

A number of indices were developed for city mapping with lichens and for comparison of air quality in different city areas. LeBlanc and DeSloover (1970) suggested Index of Atmospheric Purity (IAP). To construct IAP, one should combine number of lichen species found on the site, cover and frequency of each species, and the number of species occurring together with the given species. The higher is IAP, the lower is air pollution level. LeBlanc and DeSloover (1970) studied lichens in Montreal, Canada. They divided the city into squares of the same size, in each square they selected tree-phorophytes of nearly the same size, not shaded and having approximately similar physical and chemical bark properties to minimize influence of factors other than the target one, namely other than air pollution. Characteristics of the above mentioned lichen community and of each lichen species were determined for all trees. IAP was constructed, than five zones with different air pollution level were outlined in Montreal. Later on, IAP was modified, and it was applied in hundreds of studies aimed to accessed air quality in cities all over the world: Grenoble, France (Gombert et al. 2004), Kiev, Ukraine (Dymytrova 2009), Cincinnati, Ohio, USA (Washburn and Culley 2006), Córdoba, Argentina (Estrabou et al. 2011) to mention a few.

Other indices were developed applying similar technique and used in different countries: LGW (Luftgűte-Index) and LGI in Germany (Kricke and Loppi 2002; Kirschbaum et al. 2012), IP (Index of Poleotolerance) in Estonia (Trass 1973), LBI (Lichen Biodiversity Index) in Italy (Badin and Nimis 1996), TDI (Trend Detection Index) in Russia (Insarov et al 1999), and some others. Values of all indices are calculated for specific climatic and microclimatic conditions, and value comparison would be correct just within a narrow range of these conditions.

3.3 Lichens as Monitors of Urban Air Quality Trends

The most common lichen substrate in cities is tree bark because epiphytic lichens are not influenced by trampled down as result of recreational activity. Trees of the same species usually can be found in different parts of the city, and lichen communities growing on homogenous substrate can be compared. Most of lichen species are substrate-specific, for instance some species called acidophytes are confined to acid tree bark, and nitrophytes are confined to sub-neutral eutrophic bark.

At the last decades of twenty century, sulphur dioxide concentrations were reduced in Western Europe and North America, thus SO_2 yearly average concentration in London decreased from 350 $\mu g \cdot m^{-3}$ in 1970s to 3 $\mu g \cdot m^{-3}$ in 2001 (Bell et al. 2004). This reduction was followed by lichen re-colonization of the tree bark, by lichen desert shrinkage and by increase of lichen species diversity in small and large cities, for instance in London (Davies et al. 2007; Larsen et al. 2007), Skawina, Poland (Lisowska 2011), Grenoble area (Gombert et al. 2006), Tampere, SW Finland (Ranta 2001), in a number of cities in Italy: Turin (Isocrono et al. 2007), Rome (Munzi et al. 2007), Siena (Loppi et al. 2002), Arezzo (Loppi et al. 2003), Pistoia (Loppi and Corsini 2003), Montecatini Terme (Loppi et al. 2004), and in others.

Species composition of lichen communities arising after SO_2 reduction, nevertheless differ from that was before air pollution appeared. New lichen community is formed by a few factors including gaseous composition shift. Nowadays, the main air pollution is emitted from vehicles at ground level. Transport-related pollution influences the lichen species composition, cover and frequency. Vehicles emit a mixture of the pollutants: nitrogen oxides, carbon monoxide, carbon dioxide, volatile organic compounds, polycyclic aromatic hydrocarbons, particulates, metals and ammonia produced by catalytic converters (Baum et al. 2001). Nitrogen compounds and dust increase tree bark eutrophication and pH (Fig. 1). As result of substrate conditions change, lichen communities change as well. Acidophytic species are gradually replaced by nitrophytic ones. For instance, in London this shift was shown by Davies et al. (2007) and Larsen et al. (2007). Authors employed different quantitative recording techniques on different tree species, however results were similar: both studies provide evidence for a NO_x and other transport-related pollution influence on the composition and frequency of lichens. Herbarium records provided valuable information on previous assemblages testifying to environmental change (Purvis 2007). Ten-point scale of sensitivity to nitrogen oxides for lichens growing on European ash (*Fraxinus excelsior* L.) trees with neutral bark was suggested for London (Davies et al. 2007). Annual average NO_x and highest NO_x concentrations were used to rank lichen species. It was found that most acidophytic lichen species are more sensitive to nitrogen oxides than nitrophytic ones, however some acidophytes are more tolerant than nitrophytes.

3.4 Case Studies: Urban Lichens and Air Pollution

Change of urban air quality was assessed due to lichen investigation in cities of the world. Below we describe a few case studies published in last years.

In the city of Córdoba, Argentina, there is no permanent instrumental measurements of air pollution. To assess air pollution level, lichen study was conducted. Lichen frequency, cover and IAP values were estimated, and the number of species

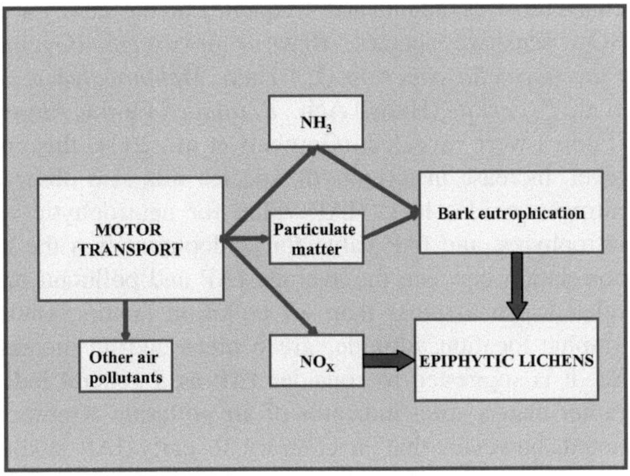

Fig. 1 Main impacts of atmospheric emission from motor transport on epiphytic lichens

and the average lichen cover were mapped. The central area of the city is a lichen desert having poor air quality. The highest IAP values and species diversity are in the southeast and northwest parts of the city. According to "poor-fair-good-very good" scale based on IAP value, the city shows fair air quality and few areas with good and very good air quality (Estrabou et al. 2011). Number of lichen species was reduced since 1998 by half as result of a huge increase in road traffic.

In tropical city San José, Costa Rica lichen cover on tree stems along the roads was correlated with traffic flow, the low traffic residential area had higher lichen covers than high traffic area. Long-term monitoring study revealed that lichen cover decreased between 1976 and 1990, and increased after 1990. Air quality amelioration possibly reflects elimination of lead from gasoline in 1989 and better traffic regulations (Monge-Nájera et al. 2002).

In the greater Cincinnati metropolitan area, USA epiphytic macrolichen diversity was studied, and comparison of urban and non-urban sites was undertaken (Washburn and Culley 2006). Average species richness per tree varied significantly between urban and non-urban sites. The highest species richness was in non-urban sites with the lowest traffic. It was demonstrated that human population density and domestic activity has less of an impact on lichen community than do mobile source emissions. Species richness, abundance and other diversity indices decline along non-urban – urban gradient.

Study of epiphytic lichens in connection to traffic-induced air pollution (characterized by distance from a road) and other human impacts was conducted in the Grenoble area, France (Gombert et al. 2004). Other human impacts were urbanization (urban, suburban or rural areas), local developments (crop fields, green areas, housing, car park slots), and exposure (trees grouped, trees isolated or in rows). Index of human impact (IHI) based on air pollution and three other human impacts was calculated for each of 345 reléve station. Reléve stations were characterized by environment types: altitude, artificiality, location of roads, industrial plants, green areas, etc. Environment types were clearly associated with lichen flora characteristics (abundance, frequency on the survey area, and species biodiversity). SO_2- sensitive species, *Bryoria fuscescens* (Gyeln.) Brodo & D. Hawksw., *Flavoparmelia caperata* (L.) Hale, *Melanohalea exasperatula* (Nyl.) O. Blanco et al., *P. perlata* (Huds.) Ach., *P. tiliacea* Flörke, *Physcia aipolia* (Ehrh. ex Humb.) (Fürnr.) were revealed, (Gombert et al., 2004) this confirms the decrease of SO_2 level. Increase in nitrophytic species was also observed, in line with increase in nitrogen oxides level. IAP value for neutrophytic species was higher than for nitrophytes, and IAP value for acidophytes was the lowest. It was no significant correlation between the average IAP and pollutant data because IAP is probably related also to other than air pollution factors: environmental artificiality, geographic location, altitude, green areas and an increase of man-made substrates. So, it is suggested to consider IAP as a general index of environmental quality rather than a strict indicator of air pollution (Gombert et al. 2004). It should be noted, however, that in contrary to early IAP studies, in Grenoble area lichen sampling was conducted at stations and trees having wide range of environmental characteristics, and this can be a reason for low correlation between the average IAP and pollution level.

Llop et al. (2012) studied epiphytic lichens in Sines, a small city on the southwest coast of Portugal. Comparison of lichen characteristics for *traffic* (areas near roads), *green* (park areas) and *house* (residential areas) was presented. Oligotrophic, hygrophytic and acidophytic lichens are sensitive to air pollution caused by traffic, and their richness near roads is less than in parks and residential areas. However, eutrophic, xerophytic and basophilous lichens are well presented in traffic areas. This can be explained by possible existence of pollutants such as alkaline dust particles and/or atmospheric NH_3, both known to cause an increase in eutrophic and nitrophytic lichens (van Herk 2001), and a decrease in oligotrophic and acidophytic ones. Overall lichen diversity decreased from *house* and *green* to *traffic*.

In Skawina, one of the major industrial centers of southern Poland, change of epiphytic lichen community on deciduous trees was studied over the last 30 years. Species richness has increased, lichen thalli vitality has improved, and former "lichen desert" in the city center has been re-colonized. While nitrogen and dust tolerant species expanded and became more frequent, some acidophytes declined (Lisowska 2011).

Our studies in Moscow, Russia also revealed change in lichen community structure during about last three decades. Since the late eightieth of twenty century, transition of air pollution sources began. Whereas until that time, industrial enterprises and power stations were the main air pollution sources, vehicle emission increased dramatically since that and constituted 92 % of total emission into atmosphere of Moscow in 2006 (Environment State 2006). Correspondingly, sulphur dioxide was replaced by nitrogen compounds and dust as main pollutants influencing lichen communities. We studied lichens response to these changes in air pollution.

The less variation of lichen community characteristics due to other than target factors (change of gaseous composition of atmosphere), the higher efficacy of lichen community trend detection. Factors masking air pollution effects are lichen community intrinsic variation and such anthropogenic factors as lichen habitat damage or destruction. To minimize variation due to these factors, one should select lichen communities for sampling within as narrow ecological stratum as possible. For epiphytic lichens such a stratum is defined by substrate (tree bark) and abiotic factors. This means that tree-phorophyte should be of the same species, nearly the same size and shading conditions. If a number of tree- phorophytes are selected, data processing should be conducted separately for each tree species, however data collected on trees with similar bark properties (pH value, roughness) can be processed jointly.

Lichens were sampled on English oak, *Quercus robur* L. having acid bark, and on lime, *Tilia* sp. having sub-neutral bark. Diameter of trees for lichen sampling was 15–38 cm at 1.5 m above the ground level, tree stems were vertical with no visible damage. There is a significant difference between lichen flora at exposed and shaded sites. Sample plots were located at woodland edge, or in parks on the assumption that tree stems at 1.5 m above the level are not shaded by undergrowth and had similar light conditions. Upon condition that trees are of the same species and of similar size, a number of trees standing close each other were selected. A desktop study was carried out to identify potential sites, and reconnaissance visits followed to ensure that adequate trees are present at each plot. This preparatory work although time consuming is important to the long-term success of the monitoring system development.

The highest possible sampling within the time available was attempted. This was achieved by surveying the maximum number of trees within the survey area over a given number of days at each site. Some plots had scarce limes and especially oaks meeting conditions described above, this was a limitation for number of model trees. On most plots we sampled lichens on 7–10 trees with vertical trunks without visible damage. Sampling plots and model trees were selected without a priori information on lichen presence/abundance at a plot and tree. All selected trees were geo-referenced.

Line-intercept method for lichen cover measurement was applied (Insarov 2002). At 1.5 m height from the tree base, a measuring tape is placed around the tree in a clockwise direction from north (zero). Cover of a lichen species at a given tree is quantified as total length of the species thalli intersections with measuring tape placed around the tree trunk at 1.5 m height from the tree base.

Comparison of relevé and line-intercept methods' efficacy was undertaken earlier. Conventional relevé method is application of sampling units of rectangular shape (e.g., Kent and Coker 1994). Measure of a method efficacy is accuracy of lichen cover estimate obtained by sampling with this method. It was proved that line-intercept method is more efficacious than relevé method (Insarov 1982). For sampling lichens on oak tree bark in European Russia, the line-intercept method is about 2.5 times more effective than the relevé method of sampling (Insarov 2002).

Twenty sample plots are established on woodland edges close to traffic flows and non-trafficked woodland edge. Lichens on 146 trees were sampled. Twenty eight lichen species were recorded on model trees, including 11 nitrophytic and five acidophytic ones. Cover and frequency of nitrophytes and acidophytes on limes and oaks were assessed. On lime trees, both cover and frequency estimates for nitrophytes significantly exceeded corresponding estimates for acidophytes. On oaks, frequency estimates for nitrophytes also exceeded corresponding estimates for acidophytes. However, acidophytes cover exceeded nitrophytes cover because of *Scoliciosporum sarothamni* (Vain.) Vězda dominance.

Increase of ammonia and dust concentrations result in eutrophication and alkalinization of the acid oak bark (Fig. 1). Lichen species diversity and community quantitative characteristic change as bark properties change. Study of epiphytic lichen communities on oak tree trunks up to 2 m above ground level in Bitsa Forest, Moscow revealed that on plots with moderate traffic influence nitrophytic lichens cover and species diversity is low. On remote from motor roads plots we found acidophytes, neutrophytes, and a few nitrophytes. Cover and diversity of nitrophytes is higher on plots near heavy traffic.

While our study was the first one in Moscow aimed to establish a base line against which lichen community change can quantified, lichen species diversity has been studied since the mid of nineteenth century. To reveal tendency of lichen species composition change, historical comparison of nitrophyte and acidophyte species diversity was undertaken. We analyzed lichen species lists for entire Moscow area within 2011 boundaries, or for its parts for period over 150 years (Insarov and Moutchnik 2007). Fraction of acidophytic lichen species decreased from 25 % in 1849 to 18 % in 2007 (found in course of our studies), while fraction of nitrophytic lichen species increased from 18 % to 39 % in the same period. Among lichen species likely disappeared since 1961, only 7 % are nitrophytes, fraction of acidophytes is as large as 41 %. These studies demonstrate that changes in epiphytic lichen community structure and composition in Moscow follow main changes in conditions of the atmospheric surface layer (Insarov et al. 2010).

Because of their physiological and ecological peculiarities, lichens are organisms combining high sensitivity to air pollution with low intrinsic variability. As a result, lichens follow main changes in gaseous compositions of the surface air. For many decades lichens were studied in cities around the world, these studies demonstrate that lichens are good monitors of air pollution in cities, and beyond. An efficient long-term lichen monitoring system can be used as an early warning system for change of urban biota stressed by air pollution.

4 Urban Planning and Management Implications

4.1 Managing Woodlands and Parks: Linking Conservation and Social Needs

It was demonstrated (e.g., Fuller et al. 2007) that in urban spaces, the degree of psychological benefit for people was positively related with green space area, habitat heterogeneity and species richness of plants and, to a lesser extent of birds and butterflies.

To avoid biotic homogenization, urban habitats should be protected and managed in the way which imitate natural processes and minimize habitat structure modification. Such management can provide social benefits of city population from one hand and proper maintenance of biodiversity from other hand (Kühn and Klotz 2006; Magura et al. 2008).

Urban green spaces are homes for small mammals, birds, ants, carabid beetles, butterflies and other animals, and they are also important components for urban dwellers. Dimension of green spaces, tree cover, structural heterogeneity of trees within forestlands are all important for support of urban fauna diversity and abundance. Moreover, it is vital to keep remains of natural landscapes including close-canopy forest patches with fallen trees, shrubs, herbs and thick litter layer. If such landscapes already disappeared, it is a challenge for urban forest planners and managers to reconstitute them. Common practice of excessive park habitat structure change by removing fallen tree trunks and leaf and branches litter should be avoided. It is also important to provide green areas connectivity within the city as well as connectivity of urban green areas with intermediate and rural landscapes.

Green areas open patches appeared as a result of wide paths and roads development are barriers for soil fauna, and as such they decrease green areas connectivity. Especially harmful are asphalt paths and roads, paved path/roads are preferable (Tóthmérész et al. 2011; Davies and Margules 1998). Planning and management of urban green areas should concurrently take into account social and economic needs and conservation purposes (Ahrné 2008; Sanesi *et al.* 2009).

4.2 Planning Roadsides: Better Air Quality

Due to physiological peculiarities, some trees are more tolerant to smoky and polluted conditions, e.g. *Pyrus calleryana* Decne (Chanticleer pear), and *Abies numidica* Glauca (Algerian fir) in the UK (Beckett et al 1998). Tree species with higher transpiration level can improve the efficiency with which particles are captured by leaf surfaces. Pollution tolerant trees should be planted as close as possible to polluted areas to absorb gaseous and particulate pollutants and thereby to improve air quality. Conifers filter pollutants year around, so they can enhance particle removal in winter when broadleaf trees do not have such an activity. Quantification of the benefits of urban trees in removing particulate pollution have been calculated e.g. by McPherson et al. (1994) who estimated that the trees of Chicago area (City of Chicago, and Cook and DuPage Counties) removed approximately 234 t of PM_{10} in 1991, improving average hourly air quality by 0.4 %. The opportunity cost of this service (based on the cost of preventing emission of the same amount of pollutant using current control strategies) was estimated at $9.2 million.

Air pollution in cities influences both the human health and biota, however it is reduced due to the vegetation. The capacity of plants to improve air quality depends on their ability to filtrate the air and to immobilize dust particles. Dust particles can be either of soil origin or produced by emission sources, e.g. vehicles on the road. Freer-Smith et al. (1997) studied accumulation of particles by oak tree leaves in Rough Wood, Walsall, UK on different distance from the road. It was found that the number of particles on oak leaves decreased with increase of distance between the tree and the road, that's why filtration efficiency of dust immobilization is maximal near the emission sources. At the same time, vegetation should not significantly reduce air exchange which can increase concentration of particles in the air near emission source. A solution is a vertical gardens with sufficient space between the plants (Litschke and Kuttler 2008). Particulate immobilization and air filtration are promising regulating service of urban roadside vegetation.

Roadside vegetation is also a good barrier for noise pollution, degree of protection depends on density, height and width of vegetation barrier. Noise is weakened owing to two mechanisms: diffusion by hedges (depends on its size and density) and resonant absorption by leaves and branches (Kowarik et al. 2011). To increase mitigation effect, it is recommended to put plants as close as possible to noise sources, including roads (Nowak and Dwyer 2007).

4.3 Long-Term Monitoring of Urban Biota with Lichens

Trend of lichen community state caused by change in air quality can be revealed earlier than that for other organisms, and for urban biota as a whole. Lichens are potential harbingers of urban biota change, and long-term monitoring is needed to realize this potentiality and to provide such a service. To ensure efficient trend detection, lichen sampling should be undertaken in narrow ecological strata. For epiphytic lichens this means that adequate number of trees with similar characteristics is available for sampling. That's why sampling should be undertaken in relatively large green areas, preferable in protected areas where construction, tree cutting and other activities lead to damage and destruction of epiphytic lichen habitats is forbidden or minimized. Lichen monitoring can become an integral part of long-term urban biota monitoring, and as such can provide information which gives policymakers more time to reach decisions provided sustainable city management and planning.

5 Concluding Remarks

Having a long history, urban ecology considerably progressed in last decades, especially in Europe. Urbanization is going on rapidly all over the world, and urban biodiversity understanding is a matter of supreme importance for conservation and for social needs, e.g. Kowarik (2011). Urbanization influences biodiversity in direct and indirect ways. Direct ways include habitat loss, habitat fragmentation and the introduction of new species. Indirect ones connected with changing urban climate, soils, hydrology, air pollution. Impacts of both types interact, affect the community structure and species composition in cities and lead to changes in urban biota comparing with natural one. Indirect impacts are usually long- term, and long term monitoring is needed to reveal biota trends caused by these impacts.

Lichen communities combine, in common, high species sensitivity to air pollution with relatively low intrinsic variability, therefore they have high signal-to-noise ratio, and this provides efficient detection of trends in lichen community state influenced by urban air pollution stress. Studies in many cities around the world demonstrate that changes in urban lichen community structure and composition follow main temporal and spatial changes in conditions of the atmospheric surface layer. Lichens proved to be effective air pollution monitors for many decades, and they can serve as harbingers of urban biota change, both in time and space, caused by air pollution and climate change stress (Sect. 3 and Nimis et al. 2002; Insarov et al. 2010; Insarov and Schroeter 2002). To realize this potential of lichens, long-term monitoring is needed.

Species richness and composition of vascular plants has also been analyzed along gradients from the urban center to the rural hinterlands of cities. In contrast to lichens (and to many groups of animal), number of plant species in cities is often higher than in rural surroundings, e.g. Knapp et al. (2009). Number of plant species and communities increase with the size of the city and the human population (e.g., Pyšek 1998). Urban floras contain considerable fraction of alien species. For instance, average proportion of alien plant species in 54 European cities appeared to be of 40 %, ranging from 20 % to 60 % (Pyšek 1998). The high species richness of urban floras can be explained by (1) high habitat heterogeneity in city areas where both natural (and semi-natural) remnants are represented along with new urban habitats like roadsides, commercial areas, etc., and (2) the high number of introduced species (Sect. 2).

In some gradient studies, e.g. in Berlin (Kowarik et al. 2011), it was found that plant species richness in the transition zone between inner city and rural areas was higher than in both ends of the gradient. Studies of other organisms also revealed species' richness peaks in peri-urban areas. An important peculiarity of peri-urban areas is high habitat heterogeneity, so they can serve as home for many species with different habitat requirements. Sprawling suburbs of the modern western cities also have a conservation value because they provide essential habitats for endemic species and facilitate their survival (Czamanski et al. 2008). Peri-urban areas of Asian mega-cities include segmented farmland patches and woodland landscapes providing key ecological functions as well as aesthetical and cultural services (Yokohari et al. 2008). Peri-urban areas can be considered as a buffer zone between cities core and agriculture areas, they can mitigate both stress for natural biota and disadvantages of city life for urban populations.

Acknowledgement Authors are grateful to an anonymous reviewer for valuable comments and linguistic advises.

References

Ahrné, K. (2008). *Local management and landscape effects on diversity of bees, wasps and birds in urban green areas*. Doctoral thesis, Swedish University of Agricultural Sciences. Uppsala: Acta Universitatis agriculturae Sueciae.

Badin, G., & Nimis, P. L. (1996). Biodiversity of epiphytic lichens and air quality in the province of Gorizia (NE Italy). *Studia Geobotanica, 15*, 73–89.

Balasooriya, B., Samson, R., Mbikwa, F., Vitharana, U. W. A., Boeckx, P., & Van Meirvenne, M. (2009). Biomonitoring of urban habitat quality by anatomical and chemical leaf characteristics. *Environmental and Experimental Botany, 65*, 386–394.

Barkman, J. J. (1958). *Phytosociology and ecology of cryptogamic epiphytes*. Assen: Van Gorcum.

Baum, M. M., Kiyomiya, E. S., Kumar, S., Lappas, A. M., Kapinus, V. A., & Lord, H. C. (2001). Multicomponent remote sensing of vehicle exhaust by dispersive absorption spectroscopy. 2. Direct on-road ammonia measurements. *Environmental Science and Technology, 35*, 3735–3741.

Beckett, K. P., Freer-Smith, P. H., & Taylor, G. (1998). Urban woodlands: Their role in reducing the effects of particulate pollution. *Environmental Pollution, 99*, 347–360.

Bell, J. N. B., Davies, L., & Honour, S. (2004). Air pollution research in London. In A. Klumpp (Ed.), *Urban air pollution, bioindication and environmental awareness* (pp. 3–16). Göttingen: Cuvillier Verlag.

Bell, J. N. B., Power, S. A., Jarraud, N., Agrawal, M., & Davies, C. (2011). The effects of air pollution on urban ecosystems and agriculture. *International Journal of Sustainable Development & World Ecology, 18*, 226–235.

Bevan, R. J., & Greenhalgh, G. N. (1976). *Rhytisma acerinum* as a biological indicator of pollution. *Environmental Pollution, 10*, 271–285.

Blair, R. B. (1999). Birds and butterflies along an urban gradient: Surrogate taxa for assessing biodiversity? *Ecological Applications, 1*, 164–170.

Brightman, F. H. (1982). Erasmus Darwin claims the earliest mention of lichens and air pollution (1790). *Bulletin of the British Lichen Society, 51*, 18.

Croci, S., Butet, A., Georges, A., Aguejdad, R., & Clergeau, P. (2008). Small urban woodlands as biodiversity conservation hot-spot: A multi-taxon approach. *Landscape Ecology, 23*, 1171–1186.

Czamanski, D., Benenson, I., Malkinson, D., Marinov, M., Roth, R., & Wittenberg, L. (2008). Urban sprawl and ecosystems—can nature survive? *International Review of Environmental and Resource Economics, 2*, 321–366.

Davies, K. F., & Margules, C. R. (1998). Effects of habitat fragmentation on carabid beetles: Experimental evidence. *Journal of Animal Ecology, 67*, 460–471.

Davies, L., Bates, J. W., Bell, J. N. B., James, P. W., & Purvis, O. W. (2007). Diversity and sensitivity of epiphytes to oxides of nitrogen in London. *Environmental Pollution, 146*, 299–310.

DeStefano, S., & DeGraaf, R. M. (2003). Exploring the ecology of suburban wildlife. *Frontiers in Ecology and the Environment, 1*(2), 95–101.

Dymytrova, L. (2009). Epiphytic lichens and bryophytes as indicators of air pollution in Kyiv city (Ukraine). *Folia Cryptogamica Estonica, 46*, 33–44.

Environment State and Nature Conservation in Russia in 2005. (2006). Moscow: Ministry of Natural Resources and Environment of the Russian Federation. In Russian, http://www.mnr.gov.ru/regulatory/list.php?part=1270. Accessed 04 Oct 2012.

Esseen, P.-A., & Renhorn, K.-A. (1998). Edge effect on an epiphytic lichen in fragmented forest. *Conservation Biology, 12*, 1307–1317.

Estrabou, C., Filippini, E., Soria, J. P., Schelotto, G., & Rodriguez, J. M. (2011). Air quality monitoring system using lichens as bioindicators in Central Argentina. *Environmental Monitoring and Assessment, 182*, 375–383.

Flückiger, W., Braun, S., & Hiltbrunner, E. (2002). Effects of air pollutants on biotic stress. In J. N. B. Bell & M. Treshow (Eds.), *Air pollution and plant life* (2nd ed., pp. 379–406). Chichester: Wiley.

Fostad, O., & Pedersen, P. A. (2000). Container-grown tree seedling responses to sodium chloride applications in different substrates. *Environmental Pollution, 109*, 203–210.

Freer-Smith, P. H., Holloway, S., & Goodman, A. (1997). The uptake of particulates by an urban woodland: Site description and particulate composition. *Environmental Pollution, 95*, 27–35.

Fuller, R. A., Irvine, K. N., Devine-Wright, P., Warren, P. H., & Gaston, K. J. (2007). Psychological benefits of greenspace increase with biodiversity. *Biology Letters, 3*, 390–394.

Gombert, S., Asta, J., & Seaward, M. R. D. (2004). Assessment of lichen diversity by index of atmospheric purity (IAP), index of human impact (IHI) and other environmental factors in an urban area (Grenoble, southeast France). *Science of the Total Environment, 324*, 183–199.

Gombert, S., Asta, J., & Seaward, M. R. D. (2006). Lichens and tobacco plants as complementary biomonitors of air pollution on the Grenoble area (Isère, southeast France). *Ecological Indicators, 6*, 429–443.

Gregg, J. W., Jones, C. G., & Dawson, T. E. (2006). Physiological and developmental effects of O_3 on cottonwood growth in urban and rural sites. *Ecological Applications, 16*, 2368–2381.

Guderian, R. (1977). *Air pollution. Phytotoxicity of acidic gases and its significance in air pollution control*. Berlin/Heidelberg/New York: Springer.

Hawksworth, D. L. (1971). Lichens as litmus for air pollution: A historical review. *International Journal of Environmental Studies, 1*, 281–296.

Hawksworth, D. L., & Rose, F. (1970). Qualitative scale for estimating sulphur dioxide air pollution in England and Wales using epiphytic lichens. *Nature, 227*, 145–148.

Hill, M. O., Roy, D. B., & Thompson, K. (2002). Hemeroby, urbanity and ruderality: Bioindicators of disturbance and human impact. *Journal of Applied Ecology, 39*, 708–720.

Hope, D., Gries, C., Zhu, W., Fagan, W. F., Redman, C. R., Grimm, N. B., Nelson, A. L., Martin, C., & Kinzig, A. (2003). Socioeconomics drive urban plant diversity. *Proceedings of the National Academy of Sciences of the United States of America, 100*, 87–92.

Iakovoglou, V., Thompson, J., & Burras, L. (2002). Characteristics of trees according to community population level and by land use in the U.S. Midwest. *Journal of Arboriculture, 28*, 59–69.

Insarov, G. (1982). Epiphytic lichen sampling on tree trunks. In *Problems of ecological monitoring and ecosystem modelling* (Vol. 5, pp. 25–33). Leningrad: Gidrometeoizdat Publisher. In Russian with English summary.

Insarov, G. (2002). A method for detecting large-scale environmental change with lichens. In P. L. Nimis, C. Scheidegger, & P. A. Wolseley (Eds.), *Monitoring with lichens – monitoring lichens* (pp. 399–403). Dordrecht: Kluwer Academic.

Insarov, G. E. (2007). Epiphytic lichens monitoring in context of air pollution/acid deposition studies of forest ecosystems in East Asian region. In *Sub-manual on EANET forest vegetation monitoring* (pp. 94–99). Niigata: Acid Deposition and Oxidant Research Center.

Insarov, G. E., & Moutchnik, E. E. (2007). Lichens and air pollution in Moscow. In *Problems of ecological monitoring and ecosystem modelling* (Vol. XXI, pp. 404–434). St-Petersburg: Hydrometeoizdat. In Russian with English abstract.

Insarov, G., & Schroeter, B. (2002). Lichen monitoring and climate change. In P. L. Nimis, C. Scheidegger, & P. A. Wolseley (Eds.), *Monitoring with lichens – monitoring lichens* (pp. 183–201). The Hague: Kluwer Academic.

Insarov, G., Insarova, I., & Semenov, S. (1999). A system to monitor climate change with epilithic lichens. *Environmental Monitoring and Assessment, 55*, 279–298.

Insarov, G. E., Moutchnik, E. E., & Insarova, I. D. (2010). Epiphytic lichens under air pollution stress in Moscow: Methodology for long-term monitoring. In *Problems of ecological monitoring and ecosystem modelling* (Vol. XXIII, pp. 277–296). Moscow: IGKE. In Russian with English abstract.

Insarova, I. D., Insarov, G. E., Bråkenhielm, S., Hultengren, S., Martinson, P.-O., & Semenov, S. (1992). Lichen sensitivity and air pollution. Swedish Environment Protection Agency. Report 4007 Uppsala, 72 pp.

Isocrono, D., Matteucci, E., Ferrarese, A., Pensi, E., & Piervittori, R. (2007). Lichen colonization in the city of Turin (North Italy) based on current and historical data. *Environmental Pollution, 145*, 258–265.

Izrael, Y. A., Filippova, L. M., Insarov, G. E., Semevsky, F. N., & Semenov, S. M. (1985). Background monitoring and analysis of the reasons of global changes in Biota state. In *Problems of ecological monitoring and ecosystem modelling* (Vol. 7, pp. 9–26). Leningrad: Hydrometeoizdat. In Russian, with English summary.

Jim, C. Y., & Chen, W. Y. (2008). Assessing the ecosystem service of air pollutant removal by urban trees in Guangzhou (China). *Journal of Environmental Management, 88*, 665–676.

Kardel, F., Wuyts, K., Babanezhad, M., Vitharana, U. W. A., Wuytack, T., Potters, G., & Samson, R. (2010). Assessing urban habitat quality based on specific leaf area and stomatal characteristics of *Plantago lanceolata* L. *Environmental Pollution, 158*, 788–794.

Kardel, F., Wuyts, K., Babanezhad, M., Wuytack, T., Adriaenssens, S., & Samson, R. (2012). Tree leaf wettability as passive bio-indicator of urban habitat quality. *Environmental and Experimental Botany, 75*, 277–285.

Kato, M., Iwata, T., Ishii, N., Hino, K., Tsutsumi, J., Nakamatsu, R., Nishime, Y., Miyagi, K., Suzuki, M., et al. (2013). In M. Kawakami (Ed.), *Spatial planning and sustainable development: Approaches for achieving sustainable urban form in Asian Cities, strategies for sustainability. Chapter 15. Effects of green curtains to improve the living environment* (pp. 271–286). Dordrecht: Springer Science + Business Media.

Kent, M., & Coker, P. (1994). *Vegetation description and analysis* (p. 363). Chichester/New York/Brisbane/Toronto/Singapore: Willey.

Kirschbaum, U., Cezanne, R., Eichler, M., Hanewald, K., & Windisch, U. (2012). Long-term monitoring of environmental change in German towns through the use of lichens as biological indicators: Comparison between the surveys of 1970, 1980, 1985, 1995, 2005 and 2010 in Wetzlar and Giessen. *Environmental Sciences Europe, 24*, 1–19. http://www.enveurope.com/content/24/1/19. Accessed 03 Mar 2013.

Knapp, S., Kühn, I., Schweiger, O., & Klotz, S. (2008). Challenging urban species diversity: Contrasting phylogenetic patterns across plant functional groups in Germany. *Ecology Letters, 11*, 1054–1064.

Knapp, S., Kühn, I., Bakker, J. P., Kleyer, M., Klotz, S., Ozinga, W. A., Poschlod, P., Thompson, K., Thuiller, W., & Romermann, C. (2009). How species traits and affinity to urban land use control large-scale species frequency. *Diversity and Distributions, 15*, 533–546.

Kowarik, I. (2011). Novel urban ecosystems, biodiversity, and conservation. *Environmental Pollution, 159*, 1974–1983.

Kowarik, I., Fischer, L. K., Säumel, I., Westermann, J. R., et al. (2011). Plants in urban settings: From patterns to mechanisms and ecosystem services. In W. Endlicher (Ed.), *Perspectives in urban ecology* (pp. 135–166). Berlin/Heidelberg: Springer.

Kricke, R., & Loppi, S. (2002). Bioindication: The I.A.P. approach. In P. L. Nimis, C. Scheidegger, & P. A. Wolseley (Eds.), *Monitoring with lichens – monitoring lichens* (pp. 21–37). Dordrecht: Kluwer Academic.

Kühn, I., & Klotz, S. (2006). Urbanization and homogenization – comparing the floras of urban and rural areas in Germany. *Biological Conservation, 127*, 292–300.

Kühn, I., Brandl, R., & Klotz, S. (2004). The flora of German cities is naturally species rich. *Evolutionary Ecology Research, 6*, 749–764.

Kunina, I. M., Insarova, I. D., & Trushin, S. B. (1979). The effect of SO_2 on metabolism of plant cell. In *Problems of ecological monitoring and ecosystem modelling* (Vol. II, pp. 87–124). Leningrad: Hydrometeoizdat. In Russian.

Larcher, W. (2003). *Physiological plant ecology: Ecophysiology and stress physiology of functional groups* (4th ed., pp. 437–450). Berlin/Heidelberg: Springer.

Larsen, R. S., Bell, J. N. B., James, P. W., Chimonides, J., Rumsey, F. J., Tremper, A., & Purvis, W. O. (2007). Lichen and bryophyte distribution on oak in London in relation to air pollution and bark acidity. *Environmental Pollution, 146*, 332–340.

Laundon, J. R., & Waterfield, A. (2007). William Borrer's lichens in the supplement to the English botany 1826–1866. *Botanical Journal of the Linnean Society, 154*, 381–392.

LeBlanc, F., & DeSloover, J. (1970). Relation between industrialization and the distribution and growth of epiphytic lichens and mosses in Montreal. *Canadian Journal of Botany, 48*, 1485–1496.

Leveau, C. M., & Leveau, L. M. (2005). Avian community response to urbanization in the Pampean region, Argentina. *Ornitologia Neotropical, 16*, 503–510.

Lisowska, M. (2011). Lichen recolonisation in an urban-industrial area of southern Poland as a result of air quality improvement. *Environmental Monitoring and Assessment, 179*, 177–190.

Litschke, T., & Kuttler, W. (2008). On the reduction of urban particle concentration by vegetation – a review. *Meteorologische Zeitschrift, 17*, 229–240.

Llop, E., Pinho, P., Matos, P., Pereira, M., & Branquinho, C. (2012). The use of lichen functional groups as indicators of air quality in a Mediterranean urban environment. *Ecological Indicators, 13*, 215–221.

Loppi, S., & Corsini, A. (2003). Diversity of epiphytic lichens and metal contents of Parmelia caperata thalli as monitors of air pollution in town of Pistoia (C Italy). *Environmental Monitoring and Assessment, 86*, 289–301.

Loppi, S., Ivanov, D., & Boccardi, R. (2002). Biodiversity of epiphytic lichens and air pollution in the town of Siena (central Italy). *Environmental Pollution, 116*, 123–128.

Loppi, S., Bosi, A., Signorini, C., & De Dominicis, V. (2003). Lichen recolonization of *Tilia* trees in Arezzo (Tuscany, central Italy) under conditions of decreasing air pollution. *Cryptogamie Mycologie, 24*, 175–185.

Loppi, S., Frati, L., Paoli, L., Bigagli, L., Rossetti, C., Bruscoli, C., & Corsini, A. (2004). Biodiversity of epiphytic lichens and heavy metal contents of *Flavoparmelia caperata* thalli as indicators of temporal variations of air pollution in the town of Montecatini Terme (central Italy). *Science of the Total Environment, 326*, 113–122.

Luley, C. J., & Bond, J. (2002). *A plan to integrate management of urban trees into air quality planning, report to Northeast State Foresters Association* (p. 73). Kent: Davey Resource Group.

Lynch, K. (2012). Sustainable urbanisation. In F. Harris (Ed.), *Global environmental issues* (Vol. 2, pp. 203–235). Chihester: Wiley

Magura, T., Tóthmérész, B. B., & Molnar, T. (2008). A species-level comparison of occurrence patterns in carabids along an urbanisation gradient. *Landscape and Urban Planning, 86*, 134–140.

McDonnell, M., Pickett, S. T. A., & Groffman, P. (1997). Ecosystem processes along an urban-torural gradient. *Urban Ecosystems, 1*, 21–36.

McKinney, M. L. (2002). Urbanization, biodiversity, and conservation. *Bioscience, 52*, 883–890.

McPherson, E. G., Nowak, D. J., & Rowntree, R. E. (1994). Chicago's Urban Forest Ecosystem: Results of the Chicago Urban Forest Climate Project. USDA General Technical Report NE-186, 202 pp.

McPherson, G., Simpson, J., Peper, P., Xiao, Q., Pettinger, D., & Hodel, D. (2001). *Tree guidelines for Inland empire communities*. Report of western center for urban forest research and education, USDA forest service, Pacific Southwest Research Station. Web http://www.fs.fed.us/psw/programs/uesd/uep/products/2/cufr_52.pdf

Monge-Nájera, J., González, M. I., Rossi, M. R., & Méndez-Estrada, V. H. (2002). Twenty years of lichen cover change in a tropical habitat (Costa Rica) and its relation with air pollution. *Revista de Biología Tropical, 50*, 309–319.
Munzi, S., Ravera, S., & Caneva, G. (2007). Epiphytic lichens as indicators of environmental quality in Rome. *Environmental Pollution, 146*, 350–358.
Nikula, S., Vapaavuori, E., & Manninen, S. (2010). Urbanization-related changes in European aspen (*Populus tremula* L.): Leaf traits and litter decomposition. *Environmental Pollution, 158*, 2132–2142.
Nimis, P. L., Scheidegger, C., & Wolseley, P. A. (Eds.). (2002). *Monitoring with lichens – Monitoring lichens*. Dordrecht: Kluwer Academic.
Nowak, D. J., & Crane, D. E. (2002). Carbon storage and sequestration by urban trees in the USA. *Environmental Pollution, 116*, 381–389.
Nowak, D. J., & Dwyer, J. F. (2007). Understanding the benefits and costs of urban forest ecosystems. In J. E. Kuser (Ed.), *Urban and community forestry in the Northeast* (Vol. 2, pp. 25–46). New York: Springer.
Nowak, D. J., Civerolo, K. L., Rao, S. T., Sistla, G., Luley, C. J., & Crane, D. E. (2000). A modeling study of the impact of urban trees on ozone. *Atmospheric Environment, 34*, 1601–1613.
Nowak, D. J., Crane, D. E., & Stevens, J. C. (2006). Air pollution removal by urban trees and shrubs in the United States. *Urban Forestry & Urban Greening, 4*, 115–123.
Oke, T. R. (1995). The heat island of the urban boundary layer: Characteristics, causes and effects. In J. E. Cermak (Ed.), *Wind climate in cities* (pp. 81–107). Amsterdam: Kluwer Academic.
Olden, J. D. (2008). Biotic homogenization. In *ELS*. Chichester: Wiley. doi:10.1002/9780470015902.a0020471. http://www.els.net.
Pandey, S., & Nagar, P. K. (2002). Leaf surface wetness and morphological characteristics of *Valeriana jatamansi* grown under open and shade habitats. *Biologia Plantarum, 45*, 291–294.
Paoletti, E. (2009). Ozone and urban forests in Italy. *Environmental Pollution, 157*, 1506–1512.
Pickett, S. T. A., Cadenasso, M. L., Grove, J. M., Boone, C. G., Groffman, P. M., Irwin, E., Kaushal, S. S., Marshall, V., McGrath, B. P., Nilon, C. H., Pouyat, R. V., Szlavecz, K., Troy, A., & Warren, P. (2011). Urban ecological systems: Scientific foundations and a decade of progress. *Journal of Environmental Management, 92*, 331–362.
Purvis, W. (2007). Lichens in a changing pollution environment: An introduction. *Environmental Pollution, 146*, 291–292.
Purvis, W. (2010). Lichens and industrial pollution. In L. C. Batty & K. B. Hallberg (Eds.), *Ecology of industrial pollution* (pp. 41–69). Cambridge, New York, Melbourne, Madrid, Cape Town, Singapore, Sao Paulo, Delhi, Dubai, Tokyo: Cambridge University Press.
Pyšek, P. (1998). Alien and native species in Central European urban floras: A quantitative comparison. *Journal of Biogeography, 25*, 155–163.
Rango, J. (2005). Arthropod communities on creosote bush (Larrea tridentata) in desert patches of varying degrees of urbanization. *Biodiversity and Conservation, 14*, 2185–2206.
Ranta, P. (2001). Changes in urban lichen diversity after a fall in sulphur dioxide levels in the city of Tampere, SW Finland. *Annales Botanici Fennici, 38*, 295–304.
Roitman, A. A. (1989). Analysis of non-uniform hierarchical data sets and its role in ecological studies. In *Problems of ecological monitoring and ecosystem modelling* (Vol. XII, pp. 324–335). Leningrad: Hydrometeoizdat. In Russian, with English summary.
Salvati, L., & Zitti, M. (2012). Monitoring vegetation and land use quality along the rural-urban gradient in a Mediterranean region. *Applied Geography, 32*, 896–903.
Sanesi, G., Padoa-Schioppa, E., Lorusso, L., & Lafortezza, R. (2009). Avian ecological diversity as an indicator of urban forest functionality. Results from two case studies in Northern and Southern Italy. *Arboriculture and Urban Forestry, 35*, 80–86.
Saunders, P. J. W. (1966). The toxicity of sulphur dioxide to *Diplocarpon rosae* Wolf causing blackspot of roses. *Annals of Applied Biology, 58*, 103–114.

Searle, S. Y., Turnbull, M. H., Boelman, N. T., Schuster, W. S. F., Yakir, D., & Griffin, K. L. (2012). Urban environment of New York City promotes growth in northern red oak seedlings. *Tree Physiology, 32*, 389–400. doi:10.1093/treephys/tps027.

Solomon, S., Qin, D., Manning, M., Chen, Z., Marquis, M., Averyt, K. B., Tignor, M., & Miller, H. L. (Eds.). (2007). *Climate change 2007: The physical science basis. Contribution of working group I to the fourth assessment report of the intergovernmental panel on climate change*. Cambridge/New York: Cambridge University Press.

Tóthmérész, B., Mátheb, I., Balázs, E., & Magura, T. (2011). Responses of carabid beetles to urbanization in Transylvania (Romania). *Landscape and Urban Planning, 101*, 330–337.

Trass, H. (1973). Lichen sensitivity to air pollution and index of poleotolerance. *Folia Cryptog Estonica, 3*, 19–22.

Trunk Epiphytes. (2004). *Manual for integrated monitoring, UN ECE convention on long-range transboundary air pollution, international cooperative programme on integrated monitoring of air pollution effects on ecosystems, ICP IM programme centre*. Helsinki: Finnish Environment Institute. http://www.environment.fi/default.asp?contentid=21139&lan=en. Accessed 04 Oct 2012.

Tzoulas, K., Korpela, K., Venn, S., Yli-Pelkonen, V., Kazmierszak, A., & Niemelä, J. (2007). Promoting ecosystem and human health in urban areas using green infrastructure: A literature review. *Landscape and Urban Planning, 81*, 167–178.

UNDESA. (2011). *World urbanization prospects: The 2011 revision*. New York: United Nations Department of Economic and Social Affairs. http://esa.un.org/unpd/wup/index.htm. Accessed 17 Aug 2012.

US EPA. (2013). Reducing urban heat islands: Compendium and strategies. *Trees and Vegetation*, 32.

van Herk, C. M. (2001). Bark pH and susceptibility to toxic air pollutants as independent cause of changes in epiphytic lichen composition in space and time. *The Lichenologist, 33*, 419–441.

von der Lippe, M., & Kowarik, I. (2007). Long-distance dispersal of plants by vehicles as a driver of plant invasions. *Conservation Biology, 21*, 986–996.

Washburn, S., & Culley, T. (2006). Epiphytic macrolichens of the greater Cincinnati metropolitan area—part II: Distribution, diversity and urban ecology. *Bryologist, 109*, 516–526.

Wuytack, T., Verheyen, K., Wuyts, K., Kardel, F., Adriaenssens, S., & Samson, R. (2010). The potential of biomonitoring of air quality using leaf characteristics of white willow (*Salix alba* L.). *Environmental Monitoring and Assessment, 171*, 197–204.

Yokohari, M., Takeuchi, K., Watanabe, T., & Yokota, S. (2008). Beyond greenbelts and zoning: A new planning concept for the environment of Asian Mega-Cities. In J. M. Marzluff, E. Shulenberger, W. Endlicher, M. Alberti, G. Bradley, C. Ryan, U. Simon, & C. ZumBrunnen (Eds.), *Urban ecology. An international perspective on the interaction between human and nature* (pp. 783–796). New York: Springer.

FSC www.fsc.org
MIX
Papier aus verantwortungsvollen Quellen
Paper from responsible sources
FSC® C105338

If you have any concerns about our products,
you can contact us on
ProductSafety@springernature.com

In case Publisher is established outside the EU,
the EU authorized representative is:
Springer Nature Customer Service Center GmbH
Europaplatz 3, 69115 Heidelberg, Germany

Printed by Libri Plureos GmbH
in Hamburg, Germany